Und   e Energy Systems

This book is accompanied by __l__ CD-ROM(s)
Check on Issue and Return

Understanding Renewable Energy Systems

Volker Quaschning

London • Sterling, VA

First published by Earthscan in the UK and USA in 2005
Reprinted 2006

ISBN 10: 1-84407-128-6
ISBN 13: 978-1-84407-128-9

Typesetting by MapSet Ltd, Gateshead, UK
Printed and bound in the UK by Bath Press, Bath
Cover design by Paul Cooper

For a full list of publications please contact:
Earthscan
8–12 Camden High Street
London, NW1 0JH, UK
Tel: +44 (0)20 7387 8558
Fax: +44 (0)20 7387 8998
Email: earthinfo@earthscan.co.uk
Web: **www.earthscan.co.uk**

22883 Quicksilver Drive, Sterling, VA 20166-2012, USA

Earthscan is an imprint of James and James (Science Publishers) Ltd and publishes in
association with the International Institute for Environment and Development

A catalogue record for this book is available from the British Library

Library of Congress Cataloging-in-Publication Data

Quaschning, Volker, 1969–
  Understanding renewable energy systems / Volker Quaschning.
      p. cm.
  Based on the German book Regenerative Energiesysteme. 3rd ed. 2003.
  Includes bibliographical references and index.
  ISBN 1-84407-128-6 (pbk.) – ISBN 1-84407-136-7 (hardback)
  1. Renewable energy sources. I. Title.
  TJ808.Q37 2005
  333.79'4–dc22   2004022852

Contents

List of Figures and Tables

FIGURES

TABLES

List of Acronyms and Abbreviations

AC	alternating current
AM	air mass
BTU	British thermal unit
CB	conduction band
CET	Central European Time
CFCs	chlorofluorocarbons
CHP	combined heat and power
CIS	copper indium diselenide
COP	coefficient of performance
CPI	consumer price index
CVD	chemical vapour deposition
DC	direct current
EG-Si	electronic-grade silicon
EPDM	ethylene propylene diene monomer
ESTIF	European Solar Thermal Industry Federation
EVA	ethylene vinyl acetate,
FB	forbidden band
FF	fill factor
GMT	Greenwich Mean Time
GTO	gate turn off
GUT	Greenwich Universal Time
HDR	hot dry rock method
IC	integrated circuit
ICS	integral collector storage
IEA	International Energy Agency
IGBT	insulated gate bipolar transistors
IPCC	Intergovernmental Panel on Climate Change
IR	infrared reflecting
kg ce	kg coal equivalent
kg oe	kg oil equivalent
LCV	lower calorific value
LEC	levelled electricity cost
LHC	levelled heat cost
MCA	maximum credible accident
MCFC	molten carbonate fuel cell
MET	Mean European Time
MG-Si	metallurgical grade silicon
MIS	metal–insulator–semiconductor
MLT	Mean Local Time

MOSFET	metal oxide semiconductor field effect transistor
MPP	maximum power point
NaS	sodium–sulphur
NiCd	nickel–cadmium
NiMH	nickel–metal hydride
NPV	net present value
PAFC	phosphoric acid fuel cell
PE	polyethylene
PP	polypropylene
ppm	parts per million
ppmv	parts per million by volume
PR	performance ratio
PR	progress ratio (Chapter 6)
PST	Pacific Standard Time
PV	photovoltaic
PWM	pulse-width modulation
R&D	research and development
rms	root mean square
SEGS	solar electric generation system
SOC	state of charge
SOFC	solid oxide fuel cell
SOG-Si	solar grade silicon
sr	steradian
STC	standard test conditions
TIM	transparent insulation material
UCV	upper calorific value
UNEP	United Nations Environmental Programme
UNFCCC	United Nations Framework Convention on Climate Change
VB	valence band
VDEW	Vereinigung Deutscher Elektrizitätswerke
VDI	Verein Deutscher Ingenieure
WMO	World Meteorological Organisation

Preface

The destruction of the environment and global warming are among the problems first mentioned in many public opinion polls that ask what are the major problems to be solved in this century. Today's energy supply is largely responsible for the anthropogenic greenhouse effect, acid rain and other negative impacts on health and the environment. The current trend is clearly not sustainable, especially given the enormous demand for energy predicted for the future. Several energy sources, however, offer the opportunity to cover our energy demand sustainably, i.e. with almost no negative influence on health and nature. These are also called renewable energy systems, because the 'fuel' is replenished by nature.

This textbook is based on the German book *Regenerative Energiesysteme*, which was first published in 1998 and became a standard text used at German universities in courses on renewable energy. Two editions have sold out and the third edition came out in 2003.

The book is aimed mainly at students, engineers, researchers and others with technical interests wanting to obtain a basic knowledge of renewable energy production. It describes the most important technical systems for using renewable energy sources, and introduces important calculation and simulation methods for these. The main focus is on technologies with high development potentials such as solar thermal systems, photovoltaics and wind power.

When describing renewable energy subjects, one has to consider technical descriptions as well as the impact on today's energy supply or sociopolitical backgrounds. A compromise between socioeconomic and technical issues must be found when dealing with energy matters. A textbook with technical focus has the obligation to describe technologies in an objective manner. However, the author's subjective influence can never be avoided entirely. The choice of contents, methods of data presentation and even the subjects left out of the book are already based on opinions.

Therefore, this book consciously renounces separation of the technological aspects from any consequences of using the technologies, or from sociopolitical aspects. The intention is to emphasize that engineers must bear in mind the potential negative impacts of the use of developed technologies. Otherwise they must accept the heavy responsibility of allowing those impacts to occur.

Those in engineering circles are often of the opinion that the development of technology itself cannot have negative consequences. It is the use of a technology that would create such consequences. However, it is irresponsible to search for technical innovations only for the sake of improving technology. The consequences of many new or even well established technologies are very difficult to estimate in many cases. Therefore, all who are involved in the development, production and application of a technology are responsible for predicting consequences critically and warning of possible dangers in time. With the aim of acknowledging this responsibility, this book always tries to point out negative consequences besides description of facts.

From my experience as a professor in the education sector, I know that the majority of people who are interested in renewable energy technologies deals intensively with the consequences of the conventional energy supply. A linking of technical with sociopolitical contents is often desired implicitly. Therefore, this textbook does not only describe technological aspects, but also deals consciously with problems of the energy industry in Chapters 1 and 6. Here, great importance was attached to substantiating all statements with objective and up-to-date facts. This allows all readers to form their own opinion.

Interesting discussions while writing this book and the very positive feedback on the German version of this book were especially motivating for me. They have shown that problems that go beyond purely technical questions are seen as very important. These problems are often ignored because they question our way of life. Solutions are difficult but not impossible to find. Constructive discussions are the first step. I hope this book can provide a contribution to such a discussion.

Volker Quaschning
Berlin, Summer 2004

Chapter 1

Energy, Climate Change and Renewable Energy Sources

THE EXPRESSION 'ENERGY'

The expression 'energy' is often used without a great deal of thought and is applied to very different contexts. In this textbook – which only deals with technically usable types of energy, especially renewables – the physical laws describing the utilization of the energy resources will be investigated. Power is inseparably linked with energy. Since many people mix up energy, work and power, the first part of this chapter will point out differences between these and related quantities.

In general, energy is the ability of a system to cause exterior impacts, for instance a force across a distance. Input or output of work changes the energy content of a body. Energy exists in many different forms such as:

- mechanical energy
- potential energy
- kinetic energy
- thermal energy
- magnetic energy
- electrical energy
- radiation energy
- nuclear energy
- chemical energy.

According to the definition above, a litre or gallon of petrol is a potential source of energy. Petrol burned in an internal combustion engine moves a car of a given mass. The motion of the car is a type of work. Heat is another form of energy. This can be seen when observing a mobile turning in the hot air ascending from a burning candle. This motion clearly demonstrates the existing force. Wind contains energy that is able to move the blades of a rotor. Similarly, sunlight can be converted to heat, thus light is another form of energy.

The *power*:

$$P = \frac{dW}{dt} = \dot{W} \qquad (1.1)$$

Table 1.1 *Conversion Factors for Energy*

	kJ	kcal	kWh	kg ce	kg oe	m³ gas	BTU
1 kilojoule (kJ)	1	0.2388	0.000278	0.000034	0.000024	0.000032	0.94781
1 kilocalorie (kcal)	4.1868	1	0.001163	0.000143	0.0001	0.00013	3.96831
1 kilowatt-hour (kWh)	3600	860	1	0.123	0.086	0.113	3412
1 kg coal equivalent (kg ce)	29,308	7000	8.14	1	0.7	0.923	27,779
1 kg oil equivalent (kg oe)	41,868	10,000	11.63	1.428	1	1.319	39,683
1 m³ natural gas	31,736	7580	8.816	1.083	0.758	1	30,080
1 British Thermal Unit (BTU)	1.0551	0.252	0.000293	0.000036	0.000025	0.000033	1

is the first derivative of the work, W, with respect to the time, t. Thus, power describes the period of time in which the correlated work is performed. For instance, if a person lifts a sack of cement 1 metre, this is work. The work performed increases the kinetic energy of the sack. Should the person lift the sack twice as fast as before, the period of time is half. Hence the power needed is twice that of before, even if the work is the same.

The units of both energy and work according to the SI unit system are joules (J), watt seconds (Ws) or newton metres (Nm), and the unit of power is the watt (W). Besides SI units a few other units are common in the energy industry. Table 1.1 shows conversion factors for most units of energy in use today. Older literature uses antiquated units such as kilogram force metre kpm (1 kpm $= 2.72 \cdot 10^{-6}$ kWh) or erg (1 erg $= 2.78 \cdot 10^{-14}$ kWh). Physics also calculates in electronvolts (1 eV $= 4.45 \cdot 10^{-26}$ kWh). The imperial unit BTU (British Thermal Unit, 1 BTU = 1055.06 J = 0.000293071 kWh) is almost unknown outside the US and the UK. Common convention is to use SI units exclusively; this book follows this convention apart from using electronvolts when describing semiconductor properties.

Many physical quantities often vary over many orders of magnitudes; prefixes help to represent these and avoid using the unwieldy exponential notation. Table 1.2 summarizes common prefixes.

Errors often occur when working with energy or power. Units and quantities are mixed up frequently. However, wrong usage of quantities can change statements or cause misunderstandings.

For example, a journal article was published in the mid-1990s in Germany describing a private photovoltaic system with a total installed power of 2.2

Table 1.2 *Prefixes*

Prefix	Symbol	Value	Prefix	Symbol	Value
Kilo	k	10^3 (thousand)	Milli	m	10^{-3} (thousandth)
Mega	M	10^6 (million)	Micro	μ	10^{-6} (millionth)
Giga	G	10^9 (billion)	Nano	n	10^{-9} (billionth)
Tera	T	10^{12} (trillion)	Pico	p	10^{-12} (trillionth)
Peta	P	10^{15} (quadrillion)	Femto	f	10^{-15} (quadrillionth)
Exa	E	10^{18} (quintillion)	Atto	a	10^{-18} (quintillionth)

Note: Words in parentheses according to US numbering system

kW. It concluded that the compensation of €0.087 to be paid per kW for feeding into the public grid was very low. Indeed, such a subsidy would be very low: it would have been 2.2 kW · €0.087/kW = €0.19 in total because it was stated as a subsidy for installed power (unit of power = kW). Although subsidies to be paid for solar electricity were quite low at that time, no owner of a photovoltaic system in Germany got as little as a total of 20 Eurocents. The author should have quoted that the payment per kilowatt *hour* (kW*h*) for electricity fed into the grid was €0.087. Assuming that the system would feed 1650 kWh per year into the grid, the system owner would get €143.55 per year. This is 750 times more than the compensation on the power basis. This example demonstrates clearly that a missing 'h' can cause significant differences.

Physical laws state that energy can neither be produced nor destroyed or lost. Nevertheless, many people talk about energy losses or energy gains, although the *law of energy conservation* states:

> *The energy content of an isolated system remains constant. Energy can neither be destroyed nor be created from nothing; energy can transform to other types of energy or can be exchanged between different parts of the system.*

Consider petrol used for moving a car: petrol is a type of stored chemical energy that is converted in a combustion engine to thermal energy, which is transformed by the pistons into kinetic energy for the acceleration of the car. Stopping the car will not destroy this energy. It will be converted to potential energy if the car climbed a hill, or to ambient heat in the form of waste heat from the engine or frictional heat from tyres, brakes and air stream. Normally, this ambient heat cannot be used anymore. Thus, driving a car converts the usable chemical energy of petrol into worthless ambient heat energy. This energy is lost as useful energy but is not destroyed. This is often paraphrased as energy loss. Hence, 'energy loss' means converting a high quality usable type of energy to a low quality non-usable type of energy.

An example illustrating the opposite is a photovoltaic system that converts sunlight to electricity. This is often described as producing energy, which, according to the law of energy conservation, is not possible. Strictly speaking,

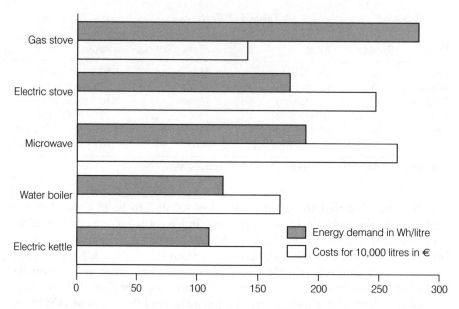

Figure 1.1 *Prices for Water Heating*

a part of the energy in the incident solar radiant energy is converted to electrical energy, i.e. the photovoltaic system converts non-usable energy to high quality energy.

Technical systems perform energy conversions with varying efficiencies. The following example should illustrate this.

The *thermal energy*, Q, which is needed to heat up one litre of water (mass $m = 1$ kg) from the temperature $\vartheta_1 = 15°C$ to $\vartheta_2 = 98°C$ is calculated with the heat capacity, c, of water $c_{H_2O} = 4187$ kJ/(kg K) using:

$$Q = c \cdot m \cdot (\vartheta_2 - \vartheta_1) \tag{1.2}$$

to $Q = 348$ kJ $= 97$ Wh.

A consumer magazine has compared different systems for boiling water. Figure 1.1 shows the results of different electrical appliances and compares them with those from a gas stove. The graph seems to show that the gas stove has the highest energy consumption while the energy costs are the lowest. The explanation is not the low price of gas, but that the graph compares different energy sources.

The electric stove uses electrical energy for water heating. Normally, this type of energy does not exist in nature, except for lightning or in electric eels. Power stations convert primary energy sources such as coal, gas or uranium into useful electricity. Conventional power stations produce large amounts of waste heat, which is emitted into the environment. They convert only a fraction of the energy stored in coal, gas or uranium into electricity, and the

Table 1.3 *Primary Energy, Final Energy and Effective Energy*

Term	Definition	Type of energy or energy source
Primary energy	Original energy, not yet processed	e.g. crude oil, coal, uranium, solar radiation, wind
Final energy	Energy in the form that reaches the end user	e.g. gas, fuel oil, petrol, electricity, hot water or steam
Effective energy	Energy in the form used by the end user	e.g. light, radiator heat, driving force of machines or vehicles

majority is 'lost'. The *efficiency*, η, describes the conversion quality and is given by:

$$efficiency \; \eta = \frac{profitable \; energy}{expended \; energy} \qquad (1.3)$$

The average thermal power station in countries such as Germany has an efficiency of around 34 per cent. Two thirds of the expended energy disappears as waste heat. This means that only one third remains as electricity.

Technical conversion of energy has different conversion stages: primary energy, final energy and effective energy. These stages are explained in Table 1.3.

Going back to the example, it has to be emphasized that the calculated thermal energy (see equation (1.2)) is the effective energy, and the values given in Figure 1.1 are final energy. The comparison of energy efficiency should, instead, be based on primary energy when considering different energy carriers such as gas and electricity. The primary energy source for generating electricity is the coal, gas or uranium used in conventional power plants. Natural gas used for boiling water is also a type of final energy. The transport of natural gas to the consumer causes some losses, but these are much lower than the

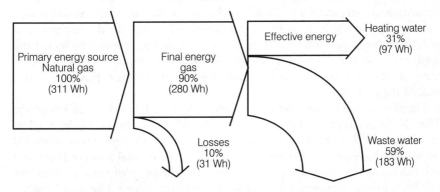

Figure 1.2 *Energy Conversion Chain and Losses for Water Heating with a Gas Cooker*

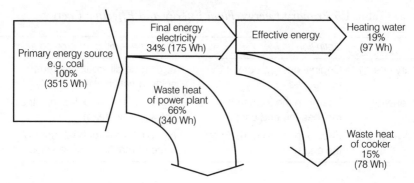

Figure 1.3 *Energy Conversion Chain and Losses for Water Heating with an Electric Cooker*

losses of the electrical transmission system (see Figure 1.2). Therefore, the primary energy consumption of the electric stove of 515 Wh = 1980 kJ is 65 per cent higher than that of the gas stove, although the final energy consumption is more than 30 per cent below that of the gas stove. This example is summarized in Figures 1.2 and 1.3, in which the energy conversion chain is compared for the electric and gas stove. The gas stove is the most economical appliance when comparing the primary energy demand, and it is the primary energy demand that determines the environmental impact.

EVOLUTION OF WORLD ENERGY DEMAND

Coal and crude oil were not relevant as energy supplies at the end of the 18th century. Firewood and techniques for using wind and hydro power provided the entire energy demand. Watermills and windmills were common features of the landscape during that time.

In 1769 James Watt laid the foundations for industrialization by developing the steam engine. The steam engine, and later the internal combustion engine, swiftly replaced mechanical wind and water installations. Coal became the single most important source of energy. In the beginning of the 20th century, crude oil took over as it was needed to support the increasing popularity of motorized road traffic. Firewood lost its importance as an energy supply in the industrial nations, and large hydro-electric power stations replaced the watermills.

The world energy demand rose sharply after the Great Depression of the 1930s. Natural gas entered the scene after World War II. In the 1960s, nuclear power was added to the array of conventional energy sources. These relatively new sources have not yet broken the predominance of coal and crude oil, but gas is the energy carrier with the fastest growth. The share of nuclear electricity of today's primary energy demand is still relatively low. The fossil energy sources – coal, crude oil and natural gas – provide more than 85 per cent of the world primary energy demand.

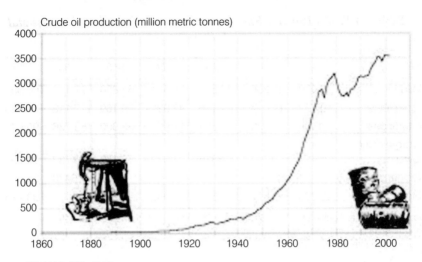

Source: BP, 2003; IEA, 2003a

Figure 1.4 *Evolution of Annual Crude Oil Production*

Figure 1.4 shows the annual oil production, illustrating the enormous increase in world energy consumption. One million metric tonnes of crude oil have an energy content of about 42 PJ or $42 \cdot 10^{15}$ J. Production rates increased exponentially after World War II. Two oil crises, in 1973 and 1978, slowed down this development, holding back the development of world economic growth and the energy demand until 1982.

Table 1.4 shows the *world primary energy consumption* of different energy sources over much of the last century. The estimation of primary energy equivalents for nuclear electricity and hydro-electricity is inconsistent; the majority of the newer statistics multiply the electricity output of nuclear power stations by 2.6 or 3 to obtain the primary energy demand. This considers the conversion efficiency of thermal power plants to be 38 per cent, or 33 per cent. The efficiency of hydro-electric power plants is much higher and can even reach values of 90 per cent or more. Since the real efficiency of hydro-electric power plants is difficult to estimate during operation, some statistics define the output as primary electricity and assume an efficiency of 100 per cent. Thus, hydro-electric power plants need much less primary energy than nuclear power plants to produce the same amount of electricity. However, statistics comparing the world primary energy supply of nuclear power plants (multiplied by 2.6 or 3) with that of hydro-electric power plants (multiplied by 1) give the impression that the hydro-electricity share is much less than that of nuclear electricity, although the world electricity supply of both is similar. Table 1.4 does not contain other renewable energy sources such as biomass (e.g. firewood and vegetable waste), wind energy, solar energy and geothermal energy. The section in this chapter (p19) on global use of renewable energy resources will describe the contribution of renewable energy.

Table 1.4 *World Primary Energy Consumption Excluding Biomass and Others*

In PJ	1925	1938	1950	1960	1968	1980	2002
Solid fuels[a]	36,039	37,856	46,675	58,541	67,830	77,118	100,395
Liquid fuels[b]	5772	11,017	21,155	43,921	79,169	117,112	147,480
Natural gas	1406	2930	7384	17,961	33,900	53,736	95,543
Hydro-electric power[c]	771	1774	3316	6632	10,179	16,732	24,792
Nuclear power[c]	0	0	0	0	463	6476	25,564
Total	43,988	53,577	78,530	127,055	191,541	271,174	393,773

Note: a Hard coal, lignite, etc.; b oil products; c converted on the basis of thermal equivalence assuming 38 per cent conversion efficiency
Source: Enquete-Kommission, 1995; BP, 2003

The global energy demand will continue to increase in the foreseeable future. It is anticipated that the increase in the industrialized nations will be lower than in developing countries, which are nonetheless catching up with the industrialized world. Furthermore, the world population is set to grow in the next few decades. Studies predict that by 2050 the energy demand will increase by a factor of 2.3 to 4 compared to 1990 (IPCC, 2000) (see also Table 1.17). This will intensify the problems of today's already high energy consumption and its consequences, such as the greenhouse effect and the rapid depletion of fossil energy resources.

The energy demand of the continents is totally different as shown in Figure 1.5. The primary energy demand of Europe, Asia and the US is certainly of the same order of magnitude. However, the population in Asia is six times that of Europe and ten times higher than that of the US. Today, the highly populated

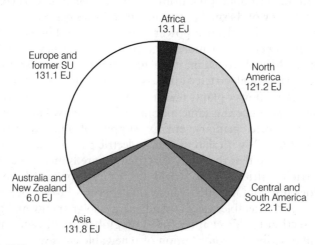

Source: DOE, 2003

Figure 1.5 *World Primary Energy Demand by Region in 2001*

Table 1.5 *Fossil Fuel Reserves*

	Crude oil	Natural gas	Coal
Proven reserves[a]	142.7 billion t $\equiv$ 5975 EJ	155.8 billion m^3 $\equiv$ 4944 EJ	984 billion t $\equiv$ 28,852 EJ
Production in 2002	3.56 billion t $\equiv$ 149 EJ	2.53 billion m^3 $\equiv$ 80 EJ	4.82 billion t $\equiv$ 141 EJ
Reserves/production ratio[a]	41 years	61 years	204 years
Unproven additional reserves[b]	84 billion t	217 billion m^3	6668 billion t[c]
Accumulated production[b]	128.2 billion t	69.6 billion m^3	–

Note: a At the end of 2002; b at the end of 2001; c total reserves;
1 t = 1 metric tonne = 2204.62 lb = 1.1023 short tonnes
Source: BP, 2003; BGR, 2002

and less-developed continents, South America and Africa, have a very small portion of the world primary energy demand. The section headed 'Greenhouse Effect' (see p10), will illustrate this uneven distribution of the energy demand by looking at the per capita carbon dioxide emissions, which correlate strongly with the energy demand.

RESERVES OF FOSSIL ENERGY SOURCES

The current energy supply depends mainly on fossil energy carriers as described in the previous section. Fossil fuels such as natural gas, petroleum, hard and brown coal needed many thousands of years to form. Organic substances (i.e. animal or vegetable residues) were the base materials. Hence, fossil fuels are stored biomass of the ancient past. A huge amount of these fossil fuels has already been consumed in the 20th century. However, due to the increasing exploitation of the fossil reservoirs, future extraction will be more and more difficult, technically challenging and risky and therefore much more expensive than today. Deep-sea oil rigs are one step in this development. If fossil fuel use continues unchecked, all available reserves of petroleum and natural gas will be exploited within the 21st century (BP, 2003). Only coal reserves will be available for a longer period of time (see Table 1.5). Thus, some decades from now, a few generations of humanity will have exploited the whole fossil energy reserves that required millions of years to form. Future generations will no longer have the opportunity to use fossil fuels as their energy supply.

An exact estimation of the existing reserves of fossil energy resources is very difficult, because only the size of deposits already explored is known. Additional reserves to be discovered in future can only be estimated. However, even if major fossil fuel reserves should be discovered, this would not change the fact that fossil fuel reserves are limited. The time span of their availability can be extended only by some years or decades at best.

Table 1.6 *Uranium (U) Resources for 2001*

	Resources with production costs		Total
	US$40/kg U	US$40–130/kg U	
Resources	1.57 Mt[a]	5.67 Mt[b]	7.24 Mt ≡ 3620 EJ
Speculative resources	12.52 Mt		12.52 Mt ≡ 6260 EJ

Note: a Reasonable assured resources; b reasonable assured resources and estimated additional resources
Source: BGR, 2002

When dealing with energy reserves, proven reserves are most important. These are reserves available with certainty, which have been proven by exploration through drilling and measurement and which are technically and economically exploitable. Additionally, unproven reserves of uncertain extent exist, but are hard to estimate. Dividing the proven reserves of an energy carrier by the present annual demand provides the statistical duration of the reserve. This duration will decrease if the energy demand rises and will increase if new reserves are exploited.

The Earth's uranium reserves for operating nuclear power stations are limited as well. The estimated global reserves are less than 20 million t, of which 12.52 million t are only speculative. Table 1.6 shows the uranium reserves. At present, only about 5 per cent of the global energy demand is provided by nuclear energy. If the total world primary energy demand in 2000, about $1.1 \cdot 10^{14}$ kWh ≈ 400 EJ, had been provided by nuclear power, the reasonably assured, economically exploitable reserves would have lasted only about 2 years. Breeder reactors can increase this time by a factor of about 60. However, nuclear power on the basis of nuclear fission is no real long-term alternative to fossil fuels due to the very restricted reserves of uranium.

Only a limited number of today's technologies will survive the 21st century due to very limited reserves of conventional energy carriers. This fact is a sufficient reason to switch our present energy supply to non-fossil and non-nuclear energy sources. This development should be completed before the conventional energy reserves are depleted. The next two sections will describe two additional motivations for a change in energy policy: the greenhouse effect and the risks of nuclear power.

GREENHOUSE EFFECT

Without the protection of Earth's atmosphere, the global mean ambient temperature would be as low as −18°C. Particular gases in the atmosphere such as carbon dioxide (CO_2), water vapour and methane capture parts of the incoming solar radiation, acting like a greenhouse. These gases have natural as well as anthropogenic, or human-induced, sources. Figure 1.6 illustrates the anthropogenic greenhouse effect.

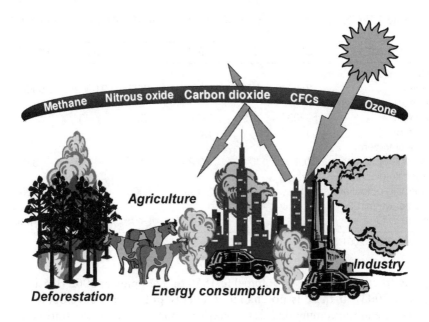

Figure 1.6 *Origin of the Anthropogenic (Human-induced) Greenhouse Effect*

The existing *natural greenhouse effect* makes life on Earth possible. Without the natural greenhouse effect, Earth would emit most of its heat radiation into space. Incident sunlight heats the Earth's surface, and the mean global ambient temperature is roughly +15°C due to the retention of this heating energy.

Over millions of years, nature has created a balance in the concentration of atmospheric gases. This has made life as we know it today possible. Several natural temperature variations have occurred over the preceding millennia, as evidenced by different ice ages.

Additional greenhouse gases are emitted to the atmosphere as a result of energy consumption and other human-induced influences. These gases cause the anthropogenic greenhouse effect. Table 1.7 summarizes the characteristics of the most important greenhouse gases.

Anthropogenic *carbon dioxide* (CO_2) results from burning fossil fuels and biomass. It contributes 61 per cent to the greenhouse effect and is the most relevant greenhouse gas. Biomass is carbon dioxide neutral if it is used at the same rate as it is grown again. On the other hand, fire clearing in the rain forest produces vast amounts of CO_2 that has been bound by these plants over decades or centuries and thus can be considered a contributor to the greenhouse effect. However, the burning of fossil fuels emits the largest amount of anthropogenic carbon dioxide. The share of fossil fuel-related carbon dioxide emissions is currently 75 per cent, and is increasing. The carbon dioxide concentration in the outer atmosphere has already risen from 280

Table 1.7 *Characteristics of Greenhouse Gases in the Atmosphere in 1998*

Greenhouse gas	CO_2	CH_4	N_2O	O_3	CFC-11	HFC-23
Concentration in ppm	365	1.745	0.314	0.03	0.000268	0.000014
Atmospheric lifetime in years	5–200	12	114	0.1	45	260
Rate of concentration change in %/year	0.4	0.4	0.25	0.5	–0.5	3.9
Specific global warming potential	1	32	150	2000	14,000	10,000
Global warming share in %	61	15	4	<9	11 (all fluorocarbons)	

Source: IPCC, 2001

ppmv (parts per million by volume) in 1850 to 372 ppmv in 2002 (Blasing and Jones, 2003). If there is no change in energy policy, this development will accelerate in the coming decades. Today's carbon dioxide concentration in the atmosphere is already higher than at any other time during the past 250,000 years.

Anthropogenic *methane* (CH_4) sources are coal mining, production of natural gas, waste disposal and agriculture such as cattle farming or cultivation of rice. The production and use of fossil fuels causes the majority of methane emissions. Although the concentration of methane in the atmosphere is less than 1 per cent of the carbon dioxide concentration, methane has a high climate change potential (15 per cent contribution to the greenhouse effect), i.e. the global warming potential of methane is much higher than that of carbon dioxide. Therefore, much smaller emission quantities are critical. In 1998 the average tropospheric concentration of 1.745 ppmv of methane had already more than doubled compared to the pre-industrial concentration of 0.7 ppmv.

Chlorofluorocarbons (CFCs) have been used in large quantities as refrigerants or propellants in spray cans. CFCs fell into disrepute mainly due to their destructive influence on the ozone layer in the stratosphere at a height of 10–50 km. International agreements to reduce CFC production in a step-by-step process initially slowed down the increase and finally decreased the concentration in recent years. However, the greenhouse potential of CFCs (11 per cent contribution to the greenhouse effect) was no significant argument in the CFC reduction discussions. Some substitutes for CFCs such as HFC-23 or R134a have significantly lower ozone-destroying potential but nearly the same greenhouse potential.

Fire clearance of tropical rain forests and the use of nitrogenous fertilizers are sources of anthropogenic *nitrous oxide* (N_2O). In 2001 the atmospheric N_2O concentration of 0.317 ppmv was only 16 per cent above the pre-industrial value; however, nitrous oxide also has a critical influence on climate change since it has a relatively long atmospheric residual time.

Motorized road traffic using fossil fuels also causes pollutants responsible for the formation of ground level *ozone* (O_3). Human-induced *stratospheric water vapour* (H_2O) also influences the greenhouse effect. However, the extent of the impact of these gases and other gases is difficult to estimate.

The contribution of different sources to the anthropogenic greenhouse effect can be summarized as:

- use of fossil fuels   50 per cent
- chemical industry (CFCs, halons)   20 per cent
- destruction of tropical rain forests
  (fire clearance, rotting)   15 per cent
- agriculture   15 per cent

These contributions vary regionally. In developing countries, the burning of rain forests and agriculture have the highest climatic influence, while in industrial countries, the use of fossil fuels dominates. As indicated in Figure 1.7, the energy demand and resulting carbon dioxide emissions also vary enormously.

Per capita, the UK emits about 9 times more carbon dioxide than does India; for Germany it is 11 times more and for the US, 20 times more. If the people of all countries caused the same amount of anthropogenic carbon dioxide as those in the US, global carbon dioxide emissions would increase five times and the anthropogenic greenhouse effect would double. The per capita CO_2 emissions of road traffic alone in Germany are twice as high as the total per capita emissions in India. In the US the picture is even worse.

The reasons for climate change are controversial. Even today, studies are published questioning the anthropogenic greenhouse effect. In fact, part of the global temperature increase of 0.6°C during the past 100 years is linked to natural fluctuations. However, the majority comes from anthropogenic emissions. In most cases it is obvious that the authors who refute the existence of anthropogenic climate change are associated with lobby groups that would be disadvantaged if a radical change of energy policy were to take place.

There are several undeniable facts substantiating a creeping climate change. In 2001, the following events were seen as indications for the increasing greenhouse effect (IPCC, 2001):

- Worldwide, 1998 was the warmest year since temperature measurements began in 1861
- The 1990s were the warmest decade since records began
- Global snow cover and the extent of ice caps has decreased by 10 per cent since the late 1960s
- Non-polar glaciers are undergoing widespread retreat
- Global mean sea level rose by 0.1–0.2 m during the 20th century
- During the 20th century, precipitation increased by 0.5–1 per cent per decade

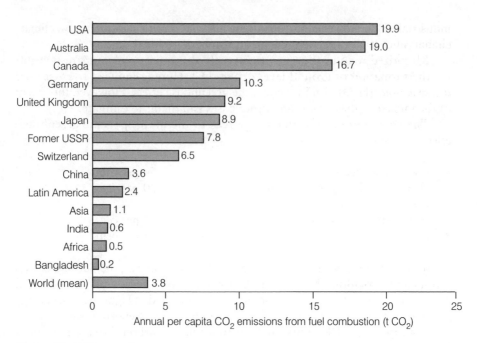

Source: IEA, 2003a

Figure 1.7 *Annual per capita Carbon Dioxide Emissions from Fuel Combustion for Different Countries in 2001*

- Heavy precipitation events increased at mid latitudes and far northern latitudes
- The frequency and intensity of droughts in Asia and Africa increased in recent decades.

A detailed prediction of the consequences of the anthropogenic greenhouse effect is not possible. Climatic models can only give an estimate of what will happen should the emission of greenhouse gases continue unchecked.

If we do not reduce anthropogenic greenhouse gas emissions, the carbon dioxide concentration in the atmosphere will more than double by the end of this century with respect to pre-industrial values. As a result, the mean global temperature will rise more than 2°C. The currently predicted range is from +1.4°C to +5.8°C. Such a temperature rise is similar to that between the ice age of 18,000 years ago and today's warm period. However, the transition from the last ice age to the current warm period took over 5000 years, but we are discussing a temperature change that will happen over 100 years.

A temperature rise of +2°C or more than +0.1°C per decade is already a very critical value, which will cause catastrophic consequences for food production and ecosystems. Global warming will have a drastic influence on global forest viability and agriculture. Globally, the lack of food in some regions will get significantly worse due to a predicted widespread decrease in agricultural production. The result could be famines and mass migrations. Furthermore, global warming will increase storm intensities with disastrous effects in tropical regions as well as in mid-range latitudes. Global sea level will rise between 0.1 and 0.9 metres in this century. The long-term sea-level rise could be as high as several metres, with severe influences on low-lying regions. Recent flood disasters give us an indication of what is to come. For example, about 139,000 people died from floods in Bangladesh in 1991. Many low-lying regions and islands will disappear from the map.

It is a sad fact that these consequences cannot be avoided completely. The limiting of global warming to +2°C is possible only if enormous efforts are undertaken immediately. By the year 2100, the various greenhouse gas levels must be reduced drastically compared with 1990 levels. This could be achieved in the following way:

- decrease global carbon dioxide emissions by 70 per cent
- reduce global N_2O and CH_4 emissions by 50 per cent and 5 per cent, respectively
- completely ban the use of all CFCs, halons and HFC-22.

Furthermore, demographers predict an increase of the world population to 15 billion (IPCC, 2000) (see also Table 1.16), more than twice the current population. Consequently, to achieve a 70 per cent reduction in global carbon dioxide emissions, the per capita carbon dioxide emissions in 2100 will have to be reduced to 15 per cent of the emissions for 1990.

The industrial countries cause the largest amount of emissions. The developing countries are currently lagging behind, but if they catch up, the industrial countries will have to achieve the following higher emission reductions for effective climate protection:

- 25 per cent reduction of CO_2 emissions by 2005 compared with 1990
- 50 per cent reduction of CO_2 emissions by 2020 compared with 1990
- 80 per cent reduction of CO_2 emissions by 2050 compared with 1990
- 90 per cent reduction of CO_2 emissions by 2100 compared with 1990.

This would mean a virtually complete withdrawal from the use of fossil fuels in this century. Technically and economically this is possible; however, in the face of a half-hearted climate policy, every possible effort has to be made.

It is not impossible to adhere to these climate reduction targets while still increasing global prosperity. However, everybody has to recognize the necessity of strong reductions and the consequences of failure. Sufficient possibilities to cover our energy demand without fossil energy sources exist today: the power industry could be based entirely on renewable energies. Thus, the question of *whether* our energy supply could be managed without fossil fuels is easy to answer. The question yet to be answered is: when will society be ready to establish a sustainable energy supply without fossil fuels and face up to its responsibilities to future generations?

NUCLEAR POWER VERSUS THE GREENHOUSE EFFECT

Nuclear fission

Since the use of fossil fuels has to be reduced significantly within the coming decades, zero-carbon energy sources are required. One option is nuclear power. Here, we distinguish between nuclear fusion and nuclear fission.

All operational nuclear power stations utilize nuclear fission for electricity generation. Here, neutrons bombard the uranium isotope ^{235}U and cause the fission of the uranium. Among others, krypton ^{90}Kr and barium ^{143}Ba are fission by-products. Furthermore, this fission reaction generates new free neutrons ^{1}n that can initiate further fission reactions. The mass of the atomic particles after the fission is reduced compared to the original uranium atom. This so-called mass defect is converted to energy ΔE in the form of heat. The following nuclear reaction equation describes the fission process:

$$^{235}_{92}U + ^{1}_{0}n \longrightarrow ^{90}_{36}Kr + ^{143}_{56}Ba + 3^{1}_{0}n + \Delta E \tag{1.4}$$

Since nature does not provide uranium in the form that is needed for technical utilization, it must be extracted from uranium ore. Rock with a uranium oxide content of more than 0.1 per cent is a workable uranium ore. Uranium mining produces huge amounts of waste that contains, in addition to some non-toxic components, a lot of radioactive residues. Uranium oxide from uranium ore contains only 0.7 per cent uranium-235. The largest portion is uranium-238, which is not usable for nuclear fission. Therefore, processing plants must enrich the uranium, i.e. the uranium-235 concentration must be increased to 2–3 per cent. In 2002, worldwide uranium production was about 34,000 metric tonnes.

Altogether, 428 nuclear power stations were in operation at the beginning of 2003, with an overall capacity of 353,505 MW. The average nuclear power station power was about 825 MW. Currently, nuclear power's share of the global primary energy demand is below 10 per cent. Figure 1.8 shows that nuclear power has a different contribution to electricity supply in different countries.

Nuclear power dominates the French electricity supply, whereas industrial nations such as Australia, Austria, Denmark, Norway or Portugal do not operate any nuclear power stations at all. Italy decided to abort nuclear power utilization after the Chernobyl disaster; Austria's decision pre-dated it. However, an electricity industry without nuclear power does not necessarily mean higher carbon dioxide emissions. For instance, hydro-electricity produces nearly 100 per cent of Norway's electricity. Iceland's electricity supply is nearly carbon dioxide free as a result of hydro and geothermal power.

If all fossil energy sources used today were replaced by nuclear power, about 10,000 new nuclear power stations would have to be built worldwide. The lifetime of a nuclear reactor is about 30 years, thus all these power stations

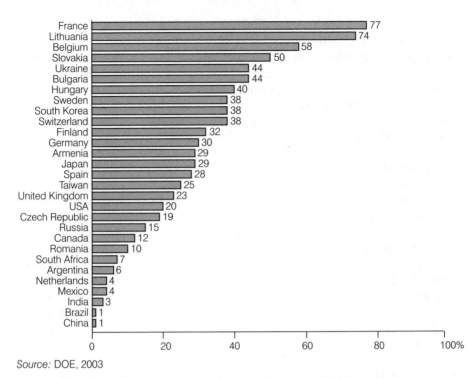

Source: DOE, 2003

Figure 1.8 *Nuclear Power's Share of Electricity Generation in 2000*

would have to be renewed after this time. Therefore, a new power plant would have to go on-line every day. In this scenario, politically unstable nations would also acquire, as an unwanted side effect, access to nuclear technology. This would increase the risk of nuclear accidents, sabotage or military use of nuclear energy, resulting in unforeseeable associated costs which make this option increasingly expensive.

As described above, Earth has limited uranium reserves. If the majority of fossil energy sources were to be replaced by nuclear power, the uranium reserves would be depleted in a short time, depending on the nuclear technology employed. Therefore, nuclear fission can only be an alternative to fossil fuels in the medium term.

Nuclear fission does not emit any carbon dioxide directly, but the building of the power plant, uranium mining, transport and disposal result in the emission of significant amounts of carbon dioxide. These indirect carbon dioxide emissions are much lower than those associated with the operation of a coal-fired power plant but higher than the indirect carbon dioxide emission of, for example, wind turbines.

Transport and storage of radioactive materials bear further risks: uranium and fuel rods must be transported to different processing plants and power stations, and radioactive waste must be transported for further treatment and to intermediate and final storage sites. Toxic and highly radioactive waste such as spent fuel rods is already produced during normal operation of a nuclear power station. Besides other radioactive substances, they also contain about 1 per cent plutonium. One microgram of plutonium, i.e. one millionth part of a gram, is considered to be lethal when breathed in; it will cause death by lung cancer. Hence, one gram of plutonium could theoretically wipe out a whole city. There is no absolute safety guarantee with such nuclear material; the possibility of transport accidents and the emission of radioactive material is very real. Final storage of radioactive waste is also very problematic, because this waste will retain its lethal properties over thousands of years.

The normal operation of nuclear power plants also bears risks. Nuclear power stations continuously emit a very small amount of radioactivity. An increased rate of leukaemia in children living near a nuclear power plant has been reported. However, accepted scientific correlations do not exist at present.

The highest risk of nuclear fission is an MCA (maximum credible accident) in a power station. Such an accident would affect millions of people and the emitted radioactivity would make large regions uninhabitable. Many humans and animals would die from radiation or fall ill with cancer. An MCA can never be totally excluded. Nuclear accidents in Harrisburg and Chernobyl have made this clear. In recent years the risk of terrorist attacks has also increased the risk of an MCA.

The first big reactor accident happened on 28 March 1979 at Three Mile Island near Harrisburg, the capital of the US state of Pennsylvania. Large amounts of radioactivity escaped. Numerous animals and plants were harmed and the number of human stillbirths in the neighbourhood of the power plant increased after the accident and cancer rates increased drastically.

On 26 April 1986, another severe nuclear reactor accident happened in the city of Chernobyl in the Ukraine, which had about 30,000 inhabitants. The escaped radioactivity not only affected the vicinity of the plant but also affected Central Europe. Many workers who tried to stop further damage at the plant paid for their efforts with their lives. The number of stillbirths and the incidence of cancer due to exposure to radiation increased significantly in the following years.

As has already been noted, civilian use of nuclear power stations is not their only potential use; they can also be used for military applications. This is one reason why civilian nuclear power has been promoted in some countries. The use of nuclear power in politically unstable countries can provoke international crises. Countries such as Iran, Iraq and North Korea have promoted nuclear power, probably also to exploit its military potential. If the use of civilian nuclear power is encouraged, the risk of nuclear crises and the risk that terrorists will come into possession of nuclear material will rise significantly.

The number of incalculable risks is balanced by the undisputed benefits of civilian use of nuclear power. Other cleaner technologies than nuclear power exist and the potentially enormous costs associated with nuclear accidents suggest that the insistence on withdrawal from the nuclear programme is perfectly justified.

Nuclear fusion

The hope of many scientists and politicians is pinned on a totally new nuclear technology: nuclear fusion. Already many billions of Euros have been spent in order to develop this technology. The model for this technology is the sun, which produces its energy by fusing hydrogen nuclei; the aim is to copy this process on Earth. Deuterium ^{2}D and tritium ^{3}T are fused to Helium ^{4}He. This reaction generates one neutron ^{1}n and energy ΔE. The equation of this reaction is:

$$\,_1^2\mathrm{D} + \,_1^3\mathrm{T} \longrightarrow \,_2^4\mathrm{He} + \,_0^1\mathrm{n} + \Delta E \tag{1.5}$$

To instigate this reaction the particles must be heated to temperatures over one million degrees Celsius. Since no known material can survive these high temperatures, technologies such as locking the reaction materials into intense magnetic fields are examined.

Deuterium and tritium for nuclear fusion exist in abundance on Earth, so this technology will not be limited by the availability of the raw materials. However, the question whether this technology will ever work cannot be answered today. Some critics say that the only thing that never changes for nuclear fusion is the remaining time needed for the development of a working reactor, which has been estimated to be 50 years for the past 50 years.

Even if this technology will work eventually, there are good reasons against spending more money on this technology. On the one hand, nuclear fusion will be much more expensive that today's nuclear fission. Renewable energies will certainly be much more economical in 50 years than nuclear fusion. On the other hand, the operation of a nuclear fusion plant also produces radioactive materials that bear risks although the risk of an MCA does not exist for a nuclear fusion plant. For the development of nuclear fusion, large amounts of money have been spent that could have been used to develop other sustainable technologies. The last and most important reason is the long time that is still needed to bring this technology to maturity. To stop global warming today, working alternatives to our present energy supply are needed urgently. Therefore, to wait for a nuclear fusion reactor that might or might not work at some time in the distant future is not the answer.

RENEWABLE ENERGIES

Even if the use of fossil fuels can be reduced significantly, and accepting that nuclear power is no long-term alternative, the question remains as to how the future supply of energy can be secured. The first step is to significantly increase the efficiency of energy usage, i.e. useful energy must be produced from a much smaller amount of primary energy, thus reducing carbon dioxide emissions.

The increasing global population and the unsatisfied demand of the developing countries will more than cancel out possible reductions due to higher energy efficiency. This problem is very well described in the book *Factor Four* by Weizsäcker and Lovins:

> *In the next 50 years, twice the current global prosperity must be achieved with half the energy and demand on natural resources* (Weizsäcker et al, 1998).

Renewable energies will be the key to this development, because they are the only option that can cover the energy demand of Earth in a climatically sustainable way.

Renewable energy sources are energy resources that are inexhaustible within the time horizon of humanity. Renewable types of energy can be subdivided into three areas: solar energy, planetary energy and geothermal energy. The respective available annual energy globally is:

- solar energy   3,900,000,000 PJ/year
- planetary energy (gravitation)   94,000 PJ/year
- geothermal energy   996,000 PJ/year

Energy stored in wind or rain, which also can be technically exploited, originate from natural energy conversions. The annual renewable energy supply exceeds the global energy demand shown in the previous section on evolution of world energy demand by several orders of magnitude. Renewable energies can, theoretically, cover the global energy demand without any problem. However, that does not mean that the transition from our present energy supply to renewable energy supplies will be possible without any problems. On the contrary, renewable energy supplies need a totally different infrastructure compared to what has been created during the past decades.

The present energy supply depends mainly on fossil energy resources. The priority is to produce and transport fossil fuels in the most economical fashion and to convert them cheaply into other types of energy in central power stations. The main advantage of fossil fuel-based energies is their ready availability. Fossil energy can be used whenever there is consumer demand.

In contrast to fossil fuels, the availability of most renewable energy sources fluctuates. A fully renewable energy supply must not only convert renewable energy sources into useful energy types, such as electricity or heat, but must also guarantee their availability. This can be done through large energy storage systems, global energy transport or adaptation of the demand to the available energy. It is not in question that renewable energies can cover our energy demand; the open questions are what share the different renewable energy sources will require for an optimal mix and when will they provide the entire energy demand. In terms of global warming, the transformation of our energy supply is the single most important challenge of the 21st century.

Geothermal energy

Geothermics utilizes the heat of Earth's interior. Geothermal power stations can utilize geothermal heat and convert it into electricity or feed it into district heating systems. In the Earth's interior, temperatures are somewhere between 3000°C and 10,000°C. Radioactive decay releases enough energy to produce such temperatures. The heat flow in the Earth's crust is relatively low and increases with depth. Volcanic eruptions demonstrate the enormous activity in the Earth's interior. The option of geothermics is demonstrated persuasively in Iceland and the Philippines, where it provides more than 20 per cent of the electricity supply.

The temperature differences between the Earth's interior and crust cause a continuous heat flow. The mean global value of this heat flow at the Earth's surface is 0.063 W/m². The total energy flow is of the same magnitude as the world's primary energy consumption. Applying the constraint that the upper strata of the Earth's crust should not be cooled down significantly and that the technical effort required must not be prohibitively high, only a portion of the available geothermal energy is usable. Today geothermal energy is only exploited in regions with geothermal anomalies. These regions record high temperatures at low depths.

Only very few regions exist with such high temperatures directly under the Earth's surface. Geysers can indicate such places. Geothermal heat pumps do not require high temperature differentials, although such differentials do help to make them more economical. Electrically driven compression heat pumps can be used to boost temperature differentials and these have reached technical maturity. However, the ecological benefits of heat pumps driven with electrical energy produced by power plants burning fossil fuels are low. On the other hand, if renewable sources provide the power required to drive the heat pump, the system becomes one avenue for providing a zero-carbon heating system. These systems are rarely used today.

Another technique that uses geothermal energy from hot, dry rocks at great depths is the so-called hot dry rock method (HDR). First a cavity is drilled into hot rocks (300°C) at a depth between 1000 and 10,000 m. Pressurized cold water is pumped into the cavity, heated up, and transported to the surface where a steam power plant generates electricity. This technology is still experimental.

Planetary energy

The different celestial bodies, in particular our moon, exchange mutual forces with Earth. The motion of the celestial bodies results in continuously varying forces at any specific point on the Earth's surface. The tides are the most obvious indicator of these forces. The movement of enormous water masses in the oceans creating the tides involves enormous amounts of energy.

Tidal energy can be used by power plants on coasts with high tidal ranges. At high tide, water is let into reservoirs and is prevented from flowing back as the tide ebbs, creating a potential difference between the collected water and water outside the reservoir. The collected water is then released though turbines into the sea at low tide. The turbines drive electric generators to produce electricity. Today there are only a few tidal power plants in operation.

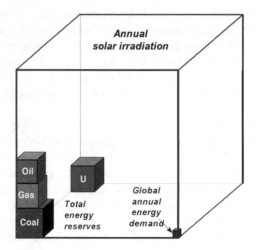

Figure 1.9 *Energy Cubes: the Annual Solar Irradiation Exceeds Several Times the Total Global Energy Demand and All Fossil Energy Reserves*

The largest tidal power plant, with a capacity of 240 MW, is situated at the Rance estuary in France. However, tidal power plants always have large impacts on nature. The amount of power that can be theoretically produced by tidal power plants globally is relatively low.

Solar energy

By far the largest energy resource is the sun. Annually, $3.9 \cdot 10^{24}$ J $= 1.08 \cdot 10^{18}$ kWh of solar energy reaches the surface of the Earth. This is about ten thousand times more than the annual global primary energy demand and much more than all available energy reserves on earth. In other words, using one-ten-thousandth part of the incoming sunlight would cover the whole energy demand of mankind. Figure 1.9 illustrates these values with energy cubes.

There is a distinction between direct and indirect solar energy. Technical systems using direct solar energy convert incoming solar radiation directly into useful energy, for instance electricity or heat. Wind, river water and plant growth are indirect forms of solar energy. Here, natural processes convert solar energy into other types of energy. Technical systems can use these indirect types of solar energy as well.

The theoretical foundations of solar radiation needed for all technical solar systems are reviewed in Chapter 2. The following sections briefly describe the different technologies using direct and indirect solar energy. Some of them are described in more detail in the following chapters.

Use of direct solar energy
The following technologies can utilize direct solar energy:

- solar thermal power plants
- photolysis systems for fuel production
- solar collectors for water heating
- passive solar heating systems
- photovoltaics, solar cells for electricity generation.

Solar thermal power plants Solar thermal power (power derived from the thermal use of solar energy) can be used to generate electricity or to produce high-temperature steam. Solar thermal power plants used for electricity generation are:

- parabolic trough power plants
- solar thermal tower power plants
- solar furnace
- Dish–Stirling systems
- solar chimney power plants.

Parabolic trough power plants were the first type of solar thermal power plant technologies operating commercially. Nine large power plants called SEGS I to IX (Solar Electric Generation System) were commissioned in California between 1984 and 1991. These power plants have a nominal capacity of between 13.8 and 80 MW each, producing 354 MW in total.

The parabolic trough collector consists of large curved mirrors, which concentrate the sunlight by a factor of 80 or more to a focal line. A series of parallel collectors are lined up in rows 300–600 metres long. Multiple parallel rows form the solar collector field. The collectors moved on one axis in order to follow the movement of the sun; this is called tracking. A collector field can also be formed by long rows of parallel Fresnel collectors. In the focal line of the collectors is a metal absorber tube, which usually is embedded into an evacuated glass tube to reduce heat losses. A special selective coating that withstands high temperatures reduces radiation heat losses.

In the Californian systems, thermo oil flows through the absorber tubes. These tubes heat the oil to 400°C. A heat exchanger transfers the thermal energy from the oil to a water–steam cycle (also called the Rankine cycle). A pump pressurizes the water and an economizer, vaporizer and a superheater jointly produce superheated steam. This steam expands in a two-stage turbine; between the high- and low-pressure parts of this turbine is a reheater. The turbine itself drives an electrical generator that converts the mechanical energy into electrical energy; the condenser after the turbine condenses the steam back to water, which allows the closing of the cycle by feeding this water back into the initial pump.

Solar collectors can also produce superheated steam directly. This makes the thermo oil superfluous and reduces costs due to savings associated with not using the expensive thermo oil. Furthermore, heat exchangers are no longer needed. However, direct solar steam generation is still at the prototype stage.

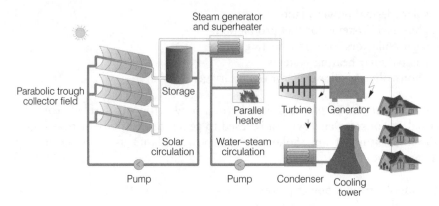

Figure 1.10 *Principle of a Parabolic Trough Solar Power Plant*

One important advantage of solar thermal power plants is that they can operate with other means of water heating and thus a hybrid system can ensure security of supply. During periods of insufficient irradiance, a parallel burner can be used to produce steam. Climate-compatible fuels such as biomass or hydrogen produced by renewable energy can also fire this parallel burner. Figure 1.10 shows the principle of a parabolic trough solar power plant.

In *solar thermal tower power plants*, hundreds, or even thousands, of large two-axis tracked mirrors are installed around a tower (Figure 1.11). These slightly curved mirrors are called heliostats. A computer is used to calculate the ideal position for each of these and positioning motors ensure precise focus on the top of the tower. The absorber is located at the focus of the mirrors on top of the tower. The absorber will be heated to temperatures of typically 1000°C or more. Hot air or molten salt transports the heat from the absorber to a steam generator where superheated steam is produced. This steam drives a turbine and an electrical generator as described above for the trough power plants. Demonstration plants exist in the US, Spain and Israel. Commercial power plants are under construction in Spain.

Figure 1.11 *Demonstration Solar Thermal Tower Power Plant in Spain*

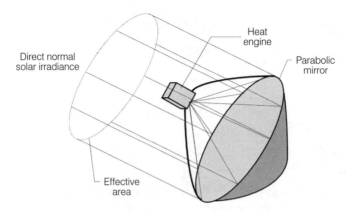

Figure 1.12 *Principle of a Dish–Stirling System*

Another system using mirrors is the solar furnace. The solar furnace in Odeillo (France) consists of various heliostat mirrors that have been set up on sloping ground. These heliostats reflect the sunlight onto a concave mirror with a diameter of 54 m. At the focus of this mirror, temperatures up to 4000°C can be reached and used for scientific experiments or, in a commercial product, for industrial processes. Further solar furnaces exist in Almería (Spain) and Cologne (Germany).

So-called Dish–Stirling systems can be used to generate electricity in the kilowatt range. A parabolic concave mirror (the dish) concentrates sunlight. A two-axis tracked mirror tracks the sun with the required high degree of accuracy. This is necessary in order to achieve high efficiencies. The receiver at the focus is heated to 650°C. The heat absorbed drives a Stirling motor, which converts the thermal energy into mechanical energy that is used to drive a generator producing electricity. If sufficient sunlight is not available, combustion heat from either fossil fuels or bio-fuels can also drive the Stirling engine and generate electricity. The system efficiency of Dish–Stirling systems can reach 20 per cent or more. Some Dish–Stirling system prototypes have been tested successfully in a number of countries; however, the cost of electricity generation using these systems is much higher than that of trough or tower power plants. Large-scale production might achieve significant cost reductions for Dish–Stirling systems. Figure 1.12 shows the principle of a Dish–Stirling system.

A *solar chimney power plant* has a high chimney (tower), with a height of up to 1000 metres. This is surrounded by a large collector roof, up to 5000 metres in diameter, that consists of glass or clear plastic supported on a framework. Towards its centre, the roof curves upwards to join the chimney, creating a funnel. The sun heats up the ground and the air under the collector roof, and the hot air follows the upward slope of the roof until it reaches the chimney. There, it flows at high speed through the chimney and drives wind generators at the bottom. The ground under the collector roof acts as thermal storage and can even heat up the air for a significant time after sunset. The

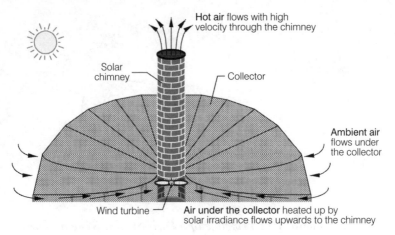

Figure 1.13 *Principle of the Solar Chimney Power Plant*

best efficiency of solar chimney power plants is currently below 2 per cent. It depends mainly on the height of the tower. Due to the large area required, these power plants can only be constructed on cheap or free land. Suitable areas could be situated in desert regions. However, the whole power plant has additional benefits, as the outer area under the collector roof can also be utilized as a greenhouse for agricultural purposes. As with trough and tower plants, the minimum economic size of a solar chimney power plant is in the multi-megawatt range. Figure 1.13 illustrates the principle of the solar chimney power plant.

Solar thermal power plants typically have poor part-load behaviour and should be installed in regions with a minimum of around 2000 full-load hours in Earth's sunbelt. However, thermal storage can increase the number of full-load hours significantly. The specific system costs are between €2000/kW and €5000/kW depending on the system size, system concept and storage size. Hence, a 50-MW solar thermal power plant will cost €100–250 million. At good sites, today's solar thermal power plants can generate electricity at around €0.15/kWh. Series production could soon bring down these costs to below €0.10/kWh. The potential for solar thermal power plants is enormous: for instance, about 1 per cent of the area of the Sahara desert covered with solar thermal power plants would theoretically be sufficient to meet the entire global electricity demand. Therefore, it is highly probable, as well as desirable, that solar thermal power systems will play an important role in the world's future electricity supply.

Solar collectors for water heating   Solar thermal energy can not only be used for the production of high-temperature heat and electricity but also for covering the demand for low-temperature heat for room heating or domestic water heating. Solar collector systems are currently used mainly for domestic water heating and swimming pool heating, but they have rarely been used for room heating systems until now.

China is by far the world's largest solar water heater manufacturer and user (see also the section on 'Use of renewable energy sources for heat and fuel protection'). By the end of 2002 the accumulated installed area of solar domestic hot water systems in China was about 40 million m^2. The annual production and sales volume was expected to reach about 8 million m^2 in 2002. There are more than one thousand manufacturers producing and selling solar thermal systems, realizing a total turnover of more than one billion Euros. Evacuated tube collectors dominate the Chinese market and exports. In Europe about 1.1 million m^2 had been installed by 2002. Here the flat-plate collector dominates. About half the new installations have been realized in Germany. Chapter 3 describes solar collectors for domestic water heating in detail.

Besides collector systems that use solar energy actively, so-called *passive* use of solar energy is possible. Well-oriented buildings with intelligently designed glass facades or transparent insulation are used for this purpose. With the combination of active and passive use of solar energy, it is also possible to build zero-energy houses, especially in moderate climatic zones such as Central Europe. The entire energy demand of these buildings is covered by solar energy. Some prototypes, for instance the energy self-sufficient solar house in Freiburg (Germany), have proven the high potential of solar energy use (Voss et al, 1995).

Photovoltaics Another technology for using solar energy to generate electricity is photovoltaics. Solar cells convert the sunlight directly into electrical energy. The potential for photovoltaic systems is huge. Even in countries with relatively low annual solar irradiance such as Germany, photovoltaics theoretically could provide more than the total electricity demand. However, an electricity supply based entirely on photovoltaic systems is not feasible since it would generate a high surplus of energy in the summer that would have to be stored at considerable cost. The combination of photovoltaic systems with other renewable energy generators such as wind power, hydro power or biomass is preferable. This would allow the whole electricity demand to be satisfied free of carbon dioxide emissions while avoiding high storage capacities and cost (Quaschning, 2001). Chapter 4 deals with photovoltaics in detail.

Indirect use of solar energy

Natural processes transform solar energy into other types of energy. Technical energy converters can utilize these indirect solar energy sources. One example for indirect types of solar energy is waterpower. Solar irradiation evaporates water from the oceans. This water rains down at higher altitudes. Streams and rivers collect this water and close the cycle at the estuaries. Hydro-electric power stations can convert the stored kinetic and potential energy of the water into electricity. Types of indirect solar energy are:

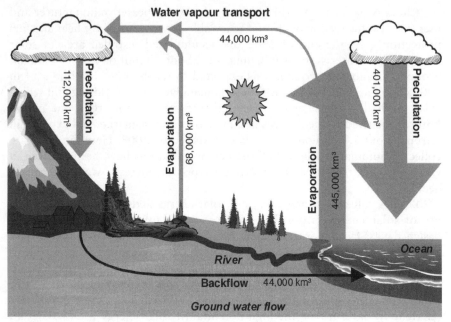

Figure 1.14 *Principle of the Global Water Cycle*

- evaporation, precipitation, water flow
- melting of snow
- wave movements
- ocean currents
- biomass production
- heating of Earth's surface and the atmosphere
- wind.

Hydro-electric power   The sun evaporates every year on average 980 litres of water from every square metre of the Earth's surface, in total 500,000 km³ (see Figure 1.14). About 22 per cent of the solar radiation energy reaching Earth is needed to drive this water cycle. Nearly 20 per cent of the evaporated water rains down on landmasses, where the majority evaporates again. About 40,000 km³ flows back to the oceans in rivers or groundwater. This is equal to more than one billion litres per second. Technically, the energy of this flow can be used.

About 160 EJ is stored in rivers and seas, which is equivalent to roughly 40 per cent of the global energy demand. About one-quarter of that energy could be technically exploited, so that nearly 10 per cent of the global energy demand could be provided free of carbon dioxide emissions by hydro-electric power. The potential for hydro-electric power in Europe is already relatively well exploited, whereas large unexploited potentials still exist in other regions of the world.

The history of using hydro power reaches back many centuries. Initially, watermills were used to convert hydro power into mechanical energy.

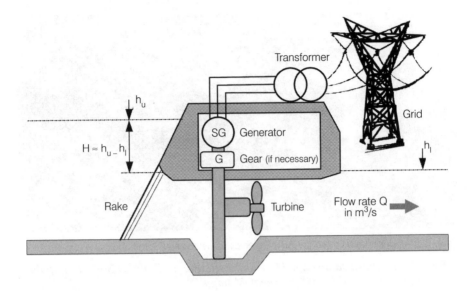

Figure 1.15 *Principle of a Hydro-electric Power Plant*

Electricity generation from hydro power started at the end of the 19th century and has achieved technical sophistication. A weir creates a height (or 'potential') difference (also called a 'head') between the water before and after the weir (see Figure 1.15). This potential difference can be utilized by a power plant. The water flows through a turbine, which transforms the potential energy into mechanical energy. An electric generator converts this into electricity. Depending on the head height and flow rate, different turbines are used. Common turbines are the Pelton, Francis or Kaplan turbines. Finally, a transformer converts the generator voltage to the grid voltage.

The power output:

$$P = \eta_G \cdot \eta_T \cdot \rho_W \cdot g \cdot Q \cdot H \tag{1.6}$$

of the power plant can be calculated from the efficiency of the generator η_G and the turbine η_T, the density ρ_W of the water ($\rho_W \approx 1000 \text{ kg/m}^3$), the head H (in m), the gravitation constant g ($g = 9.81 \text{ m/s}^2$) and the flow rate Q (in m^3/s).

Table 1.8 *Contribution of Hydro-electricity to the Net Electricity Generation in Different Countries*

Country	Paraguay	Norway	Brazil	Iceland	Venezuela	Austria	Canada
Share (%)	100	99	89	83	75	71	60
Country	Russia	China	Australia	US	Germany	UK	Netherlands
Share (%)	19	17	8	7	4	1.4	0.1

Source: DOE, 2003

Note: Left: Upper Reservoir; Right: Lower Reservoir, Penstock and Surge Shaft

Figure 1.16 *Pumped-storage Hydro-electric Power Plant in Southern Spain near Malaga*

In addition to electricity generation in river or mountain power plants there are so-called pumped-storage hydro-electric power plants (see Figure 1.16). These power plants can be used for electricity storage. In times of excess power generation, a pump transports water in a storage basin to a higher level. When the water flows back, a turbine and a generator can convert the potential energy of the water back into electricity.

Hydro-electric power is, apart from the use of biomass, the only renewable energy resource that covers a noticeable proportion of the global energy demand. The resources and the contribution of hydro-electricity to the electricity supply vary from country to country as shown in Table 1.8.

The Itaipu hydro-electric power plant shown in Figure 1.17 is situated in the border area between Brazil and Paraguay. It is at present the largest hydro-electric power plant in the world. It has a rated capacity of 12.6 GW and generated 24.3 per cent of the electricity demand of Brazil and 93.6 per cent of that of Paraguay in the year 2000. The total electricity generation in the same year was 93.4 billion kWh. Table 1.9 shows the enormous dimensions of the Itaipu power plant. However, an even larger plant is under construction: the

Table 1.9 *Technical Data of the Itaipu Hydro-electric Power Plant*

Reservoir		Dam		Generator units	
Surface	1350 km²	Maximum height	196 m	Number	18 (9 each of 50 Hz and 60 Hz)
Extent	170 km	Overall length	7760 m	Rated power	715 MW each
Volume	29,000 km³	Concrete volume	8.1 million m³	Weight	3343/3242 t each

Source: Itaipu Binacional (2003)

Source: Itaipu Binacional (2003)

Figure 1.17 *Itaipu Hydro-electric Power Plant*

Three Gorges hydro-electric power plant in China will generate 18.2 GW when completed.

Such large hydro-electric power plants are not uncontroversial because they also have a negative impact on nature and local conditions. For the Three Gorges power plant in China, several hundred thousand people had to be relocated. An example of adverse environmental effects is the Aswan dam in Egypt. It stopped the Nile from flooding, and hence from replenishing the nutrients in the intensively farmed flood planes. The artificial irrigation required to make up for the missing fertilization caused salination of the ground and harvests deteriorated. The area around the estuary also is affected by increasing soil erosion.

Before planning large hydro-electric power plants, the advantages and disadvantages should be considered carefully. On the one hand, hydro-electric power is a technology that can generate electricity without carbon dioxide emission at very low cost. On the other hand, there are negative impacts as detailed above. Small hydro-electric power plants can be an alternative. Their negative impacts are usually much smaller, but their relative costs are much higher.

The tidal power plants described on p21 also utilize hydro power. Other types of power plants that use hydro power are wave or ocean current power plants. However, these power plants are still at the prototype stage at present.

Biomass   Life on Earth is possible only because of solar energy, a substantial amount of which is utilized by plants. The following equation describes, in general, the production of biomass:

Table 1.10 *Efficiencies for Biomass Production*

Oceans	0.07%	Woods	0.55%
Fresh water	0.50%	Maize	3.2%
Man-made landscape	0.30%	Sugarcane	4.8%
Grassland	0.30%	Sugar beet	5.4%

$$H_2O + CO_2 + \text{substances} + \Delta E \longrightarrow \underbrace{C_k H_m O_n}_{\text{Biomass}} + H_2O + O_2 + \frac{\text{metabolic}}{\text{products}}$$

$$(1.7)$$

Dyes such as chlorophyll split water molecules H_2O using the energy ΔE of the visible sunlight. The hydrogen H and the carbon dioxide CO_2 taken from the air form biomass $C_k H_m O_n$. Oxygen O_2 is emitted during that process. Biomass can be used for energy in various ways. Such use converts biomass back again to CO_2 and H_2O. However, this conversion emits as much CO_2 as the plant had absorbed from the atmosphere while it was growing. Biomass is a carbon dioxide-neutral renewable energy source as long as the resource is managed sustainably.

A comparison of biomass production with other energy conversion processes is based on the estimated efficiencies of various plants. This efficiency describes what percentage of solar energy is converted to biomass. The average efficiency of global biomass production is about 0.14 per cent.

Table 1.10 shows some specific efficiencies of different methods of biomass production. The efficiency is calculated based on the calorific values given in Table 1.11 of the biomass grown in a certain area over a given time, which is then divided by the solar energy incident in this area during the same period of time.

Biomass usage can be classified as use of organic waste or agricultural residues and the cultivation of purpose-grown energy plants. Biomass can be used, for instance, in combustion engines, typically combined heat and power (CHP) plants. These CHP plants are usually smaller than large conventional coal or gas power plants, because it is important to minimize biomass transportation distances. Therefore, these power plants usually have a capacity of a few MW. Figure 1.18 shows a power station that is fired with residues from olive oil production.

Table 1.11 *Calorific Values of Various Biomass Fuels*

Fuel (anhydrous)	Lower calorific value (LCV)	Fuel (anhydrous)	LCV
Straw (wheat)	17.3 MJ/kg	China reed	17.4 MJ/kg
Non-flowering		Colza oil	37.1 MJ/kg
plants (wheat)	17.5 MJ/kg	Ethanol	26.9 MJ/kg
Wood without bark	18.5 MJ/kg	Methanol	19.5 MJ/kg
Bark	19.5 MJ/kg	Petrol (for comparison)	43.5 MJ/kg
Wood with bark	18.7 MJ/kg		

Source: Markus Maier/Steffen Ulmer

Figure 1.18 *Biomass Power Plant Using Residues of Olive Oil Production in Southern Spain*

The potential of fast-growing energy plants such as Chinese reeds or colza is very high. Even in densely populated industrial countries such as Germany, these could provide about 5 per cent of the primary energy demand without competing with food production. In the energy supply of many developing countries, biomass has a share of up to more than 90 per cent. In industrial countries, the revival of biomass use is in sight. However, some countries use more biomass than can be sustainably grown, causing enormous problems for nature. Furthermore, the cost of the technical use of biomass is usually higher than that for fossil fuels.

For the technical use of biomass there are several possibilities. Besides firing a power plant, it can be liquefied or converted to alcohol. In some regions, biomass is already used intensively as motor fuel. For instance, in Brazil, alcohol has been produced from sugar cane for some decades now. The use of bio-fuels is also increasing in other countries. The major advantage of biomass in comparison to solar or wind energy is that the stored bio-energy can be used on demand. Therefore, biomass will be an important resource to smooth fluctuations in solar and wind energy in a future climatically sustainable energy supply.

Low-temperature heat   Solar radiation heats up the surface of the Earth as well as the atmosphere. Temperature differences between the atmosphere and the Earth's surface cause compensatory air flows that are the source of wind power as described below. The Earth stores solar heat over hours, days or even months.

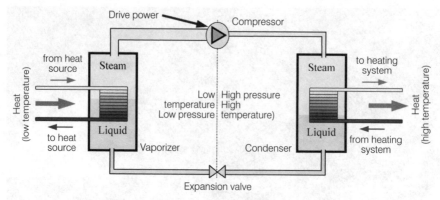

Figure 1.19 *Principle of a Compression Heat Pump*

Heat pumps can technically utilize the low-temperature heat in the ground and air as already mentioned in the section on geothermics. An exact division of such low-temperature heat into solar energy and geothermic energy is very difficult.

Figure 1.19 shows the operating principle of compression heat pumps that utilize low-temperature heat. A compressor, which is driven by an electric motor, or a gas or petrol motor powered with external energy, compresses a vaporous working medium. Pressurizing the working medium also results in significant heating. A condenser removes the heat from the working medium until it becomes a liquid. The heat at the higher temperature level is used for room or domestic water heating. The pressurized working medium expands back across an expansion valve and reaches the vaporizer. Heat from the low-temperature heat source vaporizes the working medium again and the compressor completes the cycle.

Depending on the working medium and pressure, a heat pump can even provide useful heat at high temperatures from sources with temperatures below 0°C. Ambient air, groundwater or soil can be heat sources. The mechanical energy W needed by the compressor is usually smaller than the useful heat Q_{out} provided. The ratio of both values is called coefficient of performance (COP) as defined by:

$$\varepsilon = \frac{|Q_{out}|}{W} = \frac{|\dot{Q}_{out}|}{P} \tag{1.8}$$

where P represents the electrical or mechanical power supplied to the compressor. The ideal or Carnot coefficient of performance ε_c of the heat pump is dependent on the temperature difference of the target temperature T_2 (the high temperature to the heating system) and the lower final temperature T_1:

$$\varepsilon_C = \frac{T_2}{T_2 - T_1} \tag{1.9}$$

From this equation is can be seen that low temperature differences and high temperatures of the heat source are essential for high COP values. However,

COP values are usually much lower than the ideal value. They are in the range 2.5–4 for electric heat pumps and between 1.2 and 2 for gas motor heat pumps. Low COP values destroy the ecological benefits of heat pumps. If a heat pump used for room heating (with a COP of 3) is driven by electrical energy from conventional power stations with an average efficiency of 33 per cent, the primary energy demand is the same as that of a condensing boiler with an efficiency of nearly 100 per cent. Earlier heat pumps used CFCs as the working medium, which had harmful effects on the ozone layer and the climate. Today alternative working fluids exist.

If renewable energy resources provide the mechanical power, a heat pump can produce useful heat that is climatically neutral. Therefore, the heat pump could play a more important role in a future climatically sustainable energy supply.

Wind energy   More than 100 years ago, wind power had a dominant role in the energy supply of many countries. Technically advanced windmills ground corn or pumped water. In the US thousands of Western Mills were used in agriculture, but all these windmills were mechanical systems. Wind generators providing electricity started to enter the market in the early 1980s in Denmark and the US. In Germany wind power had an unexpected boom in the 1990s, making Germany the largest wind power market in the world. In 2002, the German wind power industry achieved a turnover of nearly €4 billion and created more than 45,000 new jobs. Altogether, 15,797 wind generators with a total capacity of 15,327 MW and an electricity generation potential of 30 TWh/year were installed in Germany by mid-2004 (Ender, 2004). This is equivalent to nearly 6 per cent of Germany's electricity demand. If Germany continues with the same growth rates as in the late 1990s, it will cover more than 10 per cent of its electricity demand in a few years. More recently, Spain started a similar exploration of its wind potential.

Although Germany only has limited areas suited for setting up wind farms, the potential is considerable. Excluding conservation areas and allowing for safe distances to settlements due to noise considerations, 53.5 GW could be achieved from installations onshore. This capacity could produce 85 TWh/year, 15 per cent of the electricity demand. The German offshore potential is 23.6 GW, which could produce 79 TWh/year. The potential in other countries is even higher. In the UK, wind power could produce well above 1000 TWh/year, which is much more than the total British electricity demand. Also the US could cover its entire electricity demand using wind power. Chapter 5 describes in detail the use of wind power for electricity supply.

GLOBAL USE OF RENEWABLE ENERGY SOURCES

The contribution of renewable energy sources to the global primary energy supply was 13.5 per cent in 2001 (IEA, 2003b). Biomass is the most important renewable energy source today and has a share of 80 per cent among the renewables. Only three centuries ago, renewable energy sources provided virtually the entire global energy supply and many studies show that this might be achieved again in the future. However, other renewable energy sources such as solar power and wind power will be crucial in the future. This section shows the installed capacity and the growth rates of recent years for different renewable energy sources, starting with the electricity supply where renewables provided about 18 per cent of the global demand in 2001.

Use of renewable energy sources for electricity generation

The amount of *wind power* installed worldwide has grown rapidly since the mid-1990s. Between 1994 and 2002 the installed capacity grew nearly ten times. However, very few countries encourage this development. About 77 per cent of the global capacity in 2002 was installed in five countries. Table 1.12 shows the development of the generating capacity of wind power in these countries over recent years. Only in Denmark has wind power reached a market share of more than 10 per cent of the electricity supply. However, the potential in other countries is enormous. For instance, only 570 MW was installed in the UK in 2002, but the potential for wind power exceeds 1000 GW. In 2002 wind-generated electricity worldwide was about 50 TWh, which is 0.33 per cent of total global electricity generation. However, if the growth rates of wind power are maintained over the next two decades, wind power can become the most important renewable energy source for electricity generation, with a market share of well above 10 per cent.

The installed capacity of *photovoltaic systems* is much smaller than wind power. However, growth rates are of the same order of magnitude as wind power and the cost of photovoltaics is decreasing much faster (see Chapter 6). In 2002 about 1.6 TWh was generated by photovoltaic systems, which is equal to 0.01 per cent of global electricity generation. Table 1.13 shows the countries with the highest numbers of installed photovoltaic systems and their generating capacity between 1992 and 2002. In 2003, the growth in Japan was significantly higher than elsewhere with the result that nearly one third of global capacity is installed in Japan alone. The second largest market, Germany, contributes roughly half as much. However, the installed global capacity is about one decade behind that of wind power.

Table 1.12 *Worldwide Total Installed Wind Generator Power in GW*

Year	1994	1995	1996	1997	1998	1999	2000	2001	2002
Germany	0.60	1.09	1.55	2.08	2.88	4.44	6.11	8.73	12.00
Spain	0.08	0.13	0.25	0.51	0.88	1.81	2.84	3.55	5.04
US	1.54	1.59	1.60	1.61	2.14	2.45	2.61	4.23	4.67
Denmark	0.54	0.64	0.84	1.11	1.42	1.74	2.34	2.46	2.88
India	0.20	0.58	0.82	0.94	0.99	1.04	1.22	1.46	1.70
Others	0.55	0.75	1.05	1.38	1.85	2.46	3.34	4.48	5.77
Total	3.51	4.78	6.10	7.64	10.15	13.94	18.46	24.92	32.07

Table 1.13 *Worldwide Total Installed Photovoltaic Power in GW*

Year	1992	1994	1996	1997	1998	1999	2000	2001	2002
Japan	0.019	0.031	0.060	0.091	0.133	0.209	0.330	0.453	0.637
Germany	0.006	0.012	0.028	0.042	0.054	0.070	0.114	0.195	0.277
US	0.044	0.058	0.077	0.088	0.100	0.117	0.139	0.168	0.212
Others	0.082	0.169	0.256	0.309	0.382	0.455	0.517	0.635	0.814
Total	0.15	0.27	0.42	0.53	0.67	0.85	1.10	1.45	1.94

Source: data from IEA-PVPS, 2003 and own calculations

Table 1.14 *Worldwide Total Installed Hydro-electric Power in GW*

Year	1980	1990	1997	1998	1999	2000	2001
US	81.7	93.4	97.5	99.1	98.5	98.9	98.6
China	20.3	34.6	55.6	59.7	65.1	70.0	79.4
Canada	47.9	57.9	65.5	66.6	66.8	66.9	67.2
Brazil	27.5	44.8	53.1	54.9	56.8	59.0	61.9
Russia	a	a	44.1	43.9	43.9	43.4	43.4
Others	280.5[b]	364.4[b]	350.6	354.3	351.7	355.5	362.0
Total	457.9	595.1	666.4	678.5	682.8	693.7	712.5

Note: a No data available; b Russia included
Source: Data from DOE, 2003

Table 1.15 *Newly Installed Glazed Solar Thermal Collectors since 1990 and Total Glazed Collector Surface in Operation at the End of 2001 in 1000 m²*

	New installation							Total installation
Year	1990	1996	1997	1998	1999	2000	2001	at end of 2001
China	720	2398	2640	2882	4207	5563	6465	32,000
Japan	543	647	438	296	307	339	314	7219
Turkey	300	342	450	675	675	675	630	6422
Germany	35	269	330	350	420	620	900	3634
Israel	250	320	350	360	370	390	390	3500
Greece	204	185	197	233	185	181	175	2790
US	235	73	56	41	40	37	25	1653
Austria	40	187	182	165	141	153	160	1651

Source: data from ESTIF, 2003

In 2003 *solar thermal power plants* with a capacity of 0.354 GW were installed globally, which produce about 0.6 TWh of solar electricity per year. The potential for solar thermal power plants is enormous but the installation rates are still relatively low.

Hydro-electric power is the most important renewable energy source for electricity generation today (see Table 1.14). The growth rates are much lower than those of wind power or photovoltaics; the absolute annual increase in capacity is nonetheless rather high. With about 2650 TWh produced, nearly 17 per cent of the electricity demand was supplied by hydro-electric power plants in 2001. This share will probably decrease slightly over the coming decades because the potential for new hydro power plants is limited.

Geothermal power is also important for some national electricity markets. In Iceland, El Salvador and the Philippines, geothermal power plants produce about one fifth of the total electricity. A total of 8.3 GW has been installed worldwide and produced about 50 TWh of electricity by 2003.

Combustible *biomass* and waste are the second most important renewable energy resource for electricity generation today; however, no reliable figures about global use are available. The annual electricity production from biomass is of the order of 150 TWh.

Use of renewable energy sources for heat and fuel production

Although the potential of *solar thermal collector systems* is enormous, their contribution to the global primary energy supply is still rather low. Assuming 70 million m^2 of installed glazed collectors by the end of 2001 and an average heat production of 600 kWh/m^2, the total global solar thermal heat production was about 150 PJ. This is equal to 0.04 per cent of the global primary energy demand. China is by far the world's largest solar water heater manufacturer and user. By the end of 2002, the accumulated installed area of solar domestic hot water systems in China was about 40 million m^2. Greece had a noticeable installed area per capita of 264 m^2 per 1000 inhabitants by 2001. Projecting these per capita numbers to the global population, the total installed area would increase more than 20 times. Table 1.15 shows the number of installations of the countries with the most significant solar collector markets.

Due to its frequent non-commercial use in developing countries, solid biomass is by far the largest renewable energy source, representing 10.4 per cent of world total primary energy supply, or 77.4 per cent of global renewables supply in 2001. Africa covers about 50 per cent of its energy supply by biomass; some countries such as Mozambique or Tanzania have even higher figures of more than 90 per cent. High growth rates have been experienced by non-solid biomass combustible renewables and waste, such as renewable municipal solid waste, biogas and liquid biomass. This segment grew annually on average at 7.6 per cent since 1990. These growth rates are expected to continue even in industrial countries.

Geothermal energy is also used for heat production. However, the global share is relatively low. In the long term, *hydrogen* produced by renewable energy sources could achieve a significant share. However, since production costs of hydrogen are very high, it will take several decades to have a noticeable impact on world energy supply.

FUTURE ENERGY DEMAND AND CLIMATIC PROTECTION

The increase in the consumption of fossil energy resources is the main reason for the anthropogenic greenhouse effect. As mentioned before, the use of fossil fuels will have to be strictly limited to minimize the negative consequences of climate warming.

In the short term, however, the reverse is likely to happen: the use of fossil energy sources and the resulting carbon dioxide emissions will increase even more without immediate and drastic changes to today's global energy policy.

Evolution of the global energy demand

The Intergovernmental Panel on Climate Change (IPCC) was founded in 1988 by the World Meteorological Organisation (WMO) and the United Nations Environmental Programme (UNEP). It publishes reports that are the basis of political consultations and decisions. The IPCC has formulated several scenarios for this century, describing possible developments of energy use and greenhouse gas emissions (IPCC, 2000). These scenarios examine different developments of industrial dynamics, population growth and use of fossil fuels. The aim of this investigation is to illustrate the impact of different scenarios on the climate and to explore possible negative consequences.

The results for six scenarios, called A1F1, A1B, A1T, A2, B1 and B2, are described (see Table 1.16):

- The A1 group of scenarios shows rapid economic growth. The peak of the population size is towards the middle of the 21st century. Three technological developments form sub-scenarios. Scenario A1F1 assumes an intensive use of fossil energy resources, scenario A1B assumes a well-balanced use of all energy sources and scenario A1T assumes the forced use of carbon dioxide-free renewable energy sources.
- Scenario A2 describes heterogeneous development with continuous population increase and a slower technical development.
- In scenario B1, the population growth is similar to that of the A1 group, but it is a fast-developing, educated and service-orientated society with reductions in material intensity of production and with the introduction of clean and sustainable technologies.
- Scenario B2 assumes a continuous growth in population, but less than that of scenario A2. The emphasis of economic development is on social and ecological sustainability. New developments are accepted more slowly in this scenario than in scenario B1.

Table 1.16 and Table 1.17 show the most important assumptions made for the different scenarios in the years 2020, 2050 and 2100. The columns for 1990 and 2001 serve as reference to past development because the reference year is 1990. Table 1.18 illustrates the resulting development of carbon dioxide

Table 1.16 *Assumptions for the Evolution of World Population and Gross Domestic Product up to 2100 for Different IPCC Emission Scenarios*

Year	World population in billions					World gross domestic product in trillion US$1990				
	1990	2001	2020	2050	2100	1990	2001	2020	2050	2100
A1F1	5.3	6.1	7.6	8.7	7.1	21	30	53	164	525
A1B	5.3	6.1	7.4	8.7	7.1	21	30	56	181	529
A1T	5.3	6.1	7.6	8.7	7.0	21	30	57	187	550
A2	5.3	6.1	8.2	11.3	15.1	21	30	41	82	243
B1	5.3	6.1	7.6	8.7	7.0	21	30	53	136	328
B2	5.3	6.1	7.6	9.3	10.4	21	30	51	110	235

Source: data from IPCC, 2000; IEA, 2003a

Table 1.17 *Assumptions for the Evolution of Primary Energy Demand and Ratio of Carbon Dioxide-Free Primary Energy by 2100 for Different IPCC Emission Scenarios*

Year	Primary energy demand in EJ					Share of CO_2-free primary energy in %				
	1990	2001	2020	2050	2100	1990	2001	2020	2050	2100
A1F1	351	420	669	1431	2073	18	19	15	19	31
A1B	351	420	711	1347	2226	18	19	16	36	65
A1T	351	420	649	1213	2021	18	19	21	43	85
A2	351	420	595	971	1717	18	19	8	18	28
B1	351	420	606	813	514	18	19	21	30	52
B2	351	420	566	869	1357	18	19	18	30	49

Source: data from IPCC, 2000; IEA, 2003a

emissions for all scenarios and the impact on the climate. The relatively wide ranges for the different parameters result from uncertainties within the models used.

The implications of this scientific study are disappointing. Even the lowest impact scenario with respect to Earth's climate shows a temperature increase of 1.4°C in this century. Together with the temperature rise of about 0.6°C observed already, the expected temperature rise will be between 2°C and more than 6°C. The full impact of such a rise can hardly even be imagined today.

All scenarios assume a rapid increase in the use of renewable energy resources, even the scenarios with an intensive fossil fuel demand. The increase in carbon dioxide-free energy sources is between 400 per cent and 500 per cent. Even this enormous increase is obviously not sufficient to get the negative influences on the climate under control. Nevertheless, a carbon dioxide-free energy supply based on renewables is possible towards the end of the 21st century.

Table 1.18 *Various IPCC Emission Scenarios and Corresponding CO_2 Concentration in the Atmosphere, Average Annual Temperature Rise and Sea Level Rise by 2100*

Year	CO_2 emissions in Gt					Carbon dioxide concentration (%)	Increase by 2100 in:	
	1990	2001	2020	2050	2100		Temperature in °C	Sea level in cm
A1F1	26.0	28.7	46.5	87.6	103.4	280	3.2–5.6	18–88
A1B	26.0	28.7	46.1	60.1	49.5	210	2.1–3.8	13–70
A1T	26.0	28.7	37.7	45.1	15.8	165	1.8–3.3	12–68
A2	26.0	28.7	44.7	63.8	106.7	250	2.8–4.8	16–75
B1	26.0	28.7	38.9	41.4	15.4	150	1.4–2.6	9–57
B2	26.0	28.7	33.0	40.3	48.8	180	1.9–3.4	12–65

Source: data from IPCC, 2000; IPCC, 2001

Further possibilities for reducing carbon dioxide emissions are efficient use of energy as well as the use of lower-carbon energy sources. Table 1.19 shows that natural gas with the same energy content as coal produces only half the emissions. Since the global natural gas reserves are much lower than the coal reserves, a change to lower-carbon energy sources can only reduce emissions temporarily.

International climatic protection

International policy has recognized the urgent need of global emission reduction for protection of Earth's climate. Following long and difficult negotiations, the contracting parties of the UN Framework Convention on Climate Change (UNFCCC) agreed upon the Kyoto Protocol in 1997. It prescribes that the industrial countries listed in Annex I of the UN framework should reduce their greenhouse gas emissions by 5.2 per cent by 2012 compared to the reference year 1990. On the other hand, it can be expected that most of the developing and emerging countries will increase their emissions drastically. Therefore, the compromises of the Kyoto Protocol can only reduce the speed with which the greenhouse gas concentration is increasing in the atmosphere.

Table 1.19 *Specific CO_2 Emission Factors of Various Fuels*

Energy source	kg CO_2/kWh	kg CO_2/GJ	Energy source	kg CO_2/kWh	kg CO_2/GJ
Wood[a]	0.39	109.6	Crude oil	0.26	73.3
Peat	0.38	106.0	Kerosene	0.26	71.5
Lignite	0.36	101.2	Petrol	0.25	69.3
Hard coal	0.34	94.6	Refinery gas	0.24	66.7
Fuel oil	0.28	77.4	Liquid petroleum gas	0.23	63.1
Diesel fuel	0.27	74.1	Natural gas	0.20	56.1

Note: a Unsustainable without reforestation; sustainable wood is carbon dioxide neutral
Source: UNFCCC, 1998

Table 1.20 *Emission Limitations or Reduction Commitment Pursuant to the Kyoto Protocol and Evolution by Signatories to the Protocol*

Annex I party	Kyoto protocol targets for 2012 (%)	Greenhouse gas emissions in Mt 1990	2000	Change from 1990 to 2000 (%)
European Community	−8	4216	4072	−3.4
Liechtenstein, Monaco, Switzerland	−8	54	53	−0.8
Bulgaria, Czech Republic, Estonia, Latvia, Lithuania, Romania, Slovakia, Slovenia	−8	832	511	−38.6
US	−7	6131	7001	+14.2
Japan	−6	1247	1358	+8.9
Canada	−6	607	726	+19.6
Poland, Hungary	−6	666	471	−29.4
Croatia	−5	NA	NA	NA
New Zealand	±0	73	77	+5.2
Russian Federation	±0	3040	1965	−35
Ukraine	±0	919	455	−51
Norway	+1	52	55	+6.3
Australia	+8	425	502	+18.2
Iceland	+10	3	3	+6.9
Total	−5.2	18,265	17,250	−5.6

Note: CO_2-equivalents exclude land-use change and forestry
Source: data from UNFCCC, 2002

The Annex I countries are not necessarily required to reduce their emissions. Flexible mechanisms allow parties to trade reductions with other countries if they contribute to these reductions, for example, by giving money for conservation measures. Furthermore, natural carbon sinks such as forests can be considered. As a result of these evasive measures and the low reduction goals, there was a lot of criticism of the international agreement on climatic protection. On the other hand, without these mechanisms there would not have been any chance of ratifying the Kyoto Protocol.

Table 1.20 shows the emission limitations and reduction commitments of the contracting parties as well as the change of carbon dioxide emissions in recent years. The member countries of the European Union have very different emission goals. For instance, Germany and Denmark have to reduce their emissions by 21 per cent and the UK by 12.5 per cent, whereas Spain can increase them by 15 per cent, Greece by 25 per cent and Portugal by 27 per cent.

Economic upheaval in the former Eastern Bloc countries is the reason for the majority of emission reduction achieved to date. In coming years, an increase in greenhouse gas emissions can be expected from these countries. On the other hand, some western countries have set positive examples. The UK has replaced the bulk of its coal demand by natural gas and had reduced its emissions by 14 per cent by 1999. Luxemburg has reduced its emission by a massive 59 per cent. Both countries have already achieved their Kyoto goals for 2012.

Since 1990, the emissions of the US have shot up significantly so that it is nearly impossible for the US to achieve its Kyoto emission goal. Given this background, the US has declared it would not ratify the Kyoto Protocol. Since the US is responsible for a large amount of global greenhouse gas emissions, this is a major setback for the international efforts on climatic protection. Nevertheless, most of the other Annex I countries want to keep the Kyoto protocol alive.

The above-described scenarios reflect the present climate reality. A further reduction in CO_2 emissions by Annex I countries is not very probable, although it could be realized technically as well as economically. However, there is a glimmer of hope for the long term. Systems utilizing renewable energy resources have reached a high level of technical development. The European Union wants to drastically increase the amount of renewables used in the coming years. Should this be achieved, Europe would set an example internationally. If other countries were to follow this example, climate protection could be enhanced more than expected from today's viewpoint. The negative consequences of climate change cannot be avoided entirely, but the magnitude of the change could be limited. For this purpose, all nations have to show more responsibility in future climate negotiations and have to keep in mind that future generations will suffer the full consequences of our current irresponsible treatment of Earth's climate.

Chapter 2

Solar Radiation

INTRODUCTION

The sun is by far the most significant source of renewable energy. The previous chapter showed that energy from geothermal sources and planetary gravitation are relatively insignificant compared with solar energy. This abundant source of energy can be utilized directly by solar thermal or photovoltaic systems. In principle, wind power and hydro-electricity are transformed solar energy and are sometimes called indirect solar energy.

This chapter is dedicated to solar irradiation, since knowledge of solar irradiation is important for many renewable energy systems. Later sections contain calculations in the field of photometry. Table 2.1 summarizes the most important photometric quantities; for the use of solar energy, mainly radiation physics-based quantities are used. Daylight quantities refer only to the visible part of light, whereas solar energy also includes invisible ultraviolet and infrared light.

Many of the following calculations use physical constants, a summary of which is given in the Appendix.

THE SUN AS A FUSION REACTOR

The sun is the central point of our solar system; it has probably been in existence for 5 billion years and is expected to survive for a further 5 billion

Table 2.1 *Important Radiant Physical Quantities and Daylight Quantities*

	Radiant physical quantities			Daylight quantities		
Name	Symbol	Unit	Name	Symbol	Unit
Radiant energy	Q_e	Ws = J	Quantity of light	Q_v	lm s
Radiant flux/radiant power	Φ_e	W	Luminous flux	Φ_v	lm
Specific emission	M_e	W/m²	Luminous exitance	M_v	lm/m²
Radiant intensity	I_e	W/sr	Luminous intensity	I_v	cd = lm/sr
Radiance	L_e	W/(m² sr)	Luminance	L_v	cd/m²
Irradiance	$E_e\,G$	W/m²	Illuminance	E_v	lx = lm/m²
Irradiation	H_e	Ws/m²	Light exposure	H_v	lx s

Note: W = watt; m = metre; s = second; sr = steradian; lm = lumen; lx = lux; cd = candela
Source: DIN, 1982; ISO, 1993

Table 2.2 *Data for the Sun and the Earth*

	Sun	Earth	Ratio
Diameter (km)	1,392,520	12,756	1:109
Circumference (km)	4,373,097	40,075	1:109
Surface (km²)	$6.0874 \cdot 10^{12}$	$5.101 \cdot 10^{8}$	1:11,934
Volume (km³)	$1.4123 \cdot 10^{18}$	$1.0833 \cdot 10^{12}$	1:1,303,670
Mass (kg)	$1.9891 \cdot 10^{30}$	$5.9742 \cdot 10^{24}$	1:332,946
Average density (g/cm³)	1.409	5.516	1:0.26
Gravity (surface) (m/s²)	274.0	9.81	1:28
Surface temperature (K)	5777	288	1:367
Centre temperature (K)	15,000,000	6700	1:2200

years. The sun consists of about 80 per cent hydrogen, 20 per cent helium and only 0.1 per cent other elements. Table 2.2 contains data about the sun in comparison to the Earth.

Nuclear fusion processes create the radiant power of the sun. During these processes, four hydrogen nuclei (protons ^{1}p) fuse to form one helium nucleus (alpha particle $^4\alpha$). The alpha particle consists of two neutrons ^{1}n and two positively charged protons ^{1}p. Furthermore, this reaction produces two positrons e$^+$ and two neutrinos v_e and generates energy. The equation of the gross reaction illustrated in Figure 2.1 is:

$$4^1_1p \longrightarrow {}^4_2\alpha + 2e^+ + 2v_e + \Delta E \tag{2.1}$$

Comparing the masses of the atomic particles before and after the reaction shows that the total mass is reduced by the reaction. Table 2.3 shows the necessary particle masses for the calculation of the mass difference. The mass of the neutrinos v_e can be ignored in this calculation and the mass of a positron e$^+$ is the same as that of an electron e$^-$.

The mass difference Δm will be calculated by:

$$\Delta m = 4 \cdot m(^1p) - m(^4\alpha) - 2 \cdot m(e^+) \tag{2.2}$$

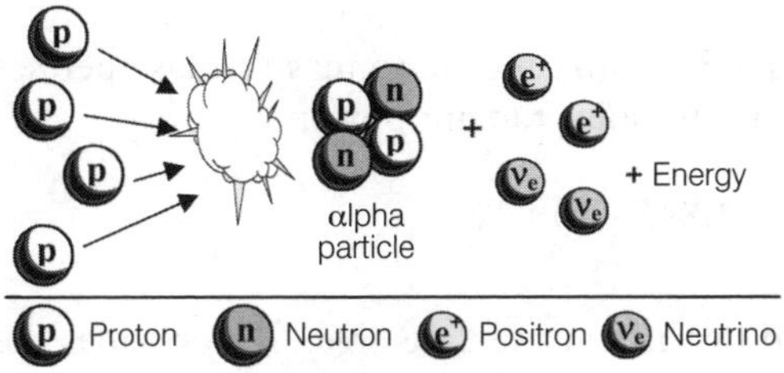

Figure 2.1 *Fusion of Four Hydrogen Nuclei to Form One Helium Nucleus (Alpha Particle)*

Table 2.3 *Various Particle and Nuclide Masses (1 u = 1.660565·10^{-27} kg)*

Particle or nuclide	Mass	Particle or nuclide	Mass
Electron (e⁻)	0.00054858 u	Hydrogen (¹H)	1.007825032 u
Proton (¹p)	1.00727647 u	Helium (⁴He)	4.002603250 u
Neutron (¹n)	1.008664923 u	Alpha particle (⁴α)	4.0015060883 u

For this reaction the result is:

$$\Delta m = 4 \cdot 1.00727647 \text{ u} - 4.0015060883 \text{ u} - 2 \cdot 0.00054858 \text{ u} = 0.02650263 \text{ u}$$

Thus, the total mass of all particles after the fusion is less than that before. The mass difference is converted into energy ΔE, with the relationship:

$$\Delta E = \Delta m \cdot c^2 \tag{2.3}$$

With the speed of light c = 2.99792458·10^8 m/s, this equation determines the energy released by the fusion as $\Delta E = 3.955 \cdot 10^{-12}$ J = 24.687 MeV. The binding energy E_b of a nucleus explains the different masses after the fusion as well as the energy difference. An atomic nucleus consists of N neutrons ¹n and Z protons ¹p. To maintain equilibrium, this binding energy has to be released during the assembly of a nucleus with protons and neutrons. The mass difference of the alpha particle and the two neutrons together with the two protons determines the binding energy of a helium nucleus.

So far, only the atomic nuclei have been considered; the electrons in the atomic shell have not been taken into account. There is one electron in the atomic shell of a hydrogen atom ¹H, while there are two electrons in the helium atom ⁴He. During the nuclear fusion process, two of the four electrons of the hydrogen atoms become the atomic shell electrons of the helium atom. The two other electrons and the positrons convert directly into energy. This radiative energy is four times the equivalent mass of an electron of 2.044 MeV. The total energy released during the reaction is thus 26.731 MeV. This very small amount of the energy does not appear to be significant at first glance; however, the enormous number of fusing nuclei results in the release of vast quantities of energy.

The sun loses 4.3 million metric tonnes of mass per second (Δm 4.3·10^9 kg/s). This results in the solar radiant power $\phi_{e,S}$ of:

$$\phi_{e,S} = \Delta m \cdot c^2 = 3.845 \cdot 10^{26} \text{ W} \tag{2.4}$$

This value divided by the sun's surface area, A_S, provides the *specific emission of the sun*:

$$M_{e,S} = \frac{\phi_{e,S}}{A_S} = 63.11 \frac{MW}{m^2} \qquad (2.5)$$

Every square metre of the sun's surface emits a radiant power of 63.11 MW. One fifth of a square kilometre of the sun's surface emits radiant energy of 400 EJ per year. This amount of energy is equal to the total primary energy demand on Earth at present.

The sun's irradiance can be approximated to that of a black body. The *Stefan–Boltzmann law*:

$$M_e (T) = \sigma \cdot T^4 \qquad (2.6)$$

can be used to estimate the *surface temperature of the sun*, T_{sun}. With the Stefan–Boltzmann constant $\sigma = 5.67051 \cdot 10^{-8}$ W/(m² K⁴), it becomes:

$$T_{sun} = \sqrt[4]{\frac{M_{e,S}}{\sigma}} = 5777 \text{ K} \qquad (2.7)$$

The surface, A_{SE}, of a sphere with the sun as its centre and with a radius equal to the average distance from the Earth to the centre of the sun ($r_{SE} = 1.5 \cdot 10^8$ km) receives the same total radiant power as the surface of the sun A_S (Figure 2.2). However, the specific emission, $M_{e,S}$, or the energy density measured over one square metre, is much higher at the sun's surface than at the sphere surrounding the sun.

With $M_{e,S} \cdot A_S = E_e \cdot A_{SE}$ and substituting $A_{SE} = 4 \cdot \pi \cdot r_{SE}^2$, the irradiance at the Earth, E_e, finally becomes:

$$E_e = M_{e,S} \cdot \frac{A_S}{A_{SE}} = M_{e,S} \cdot \frac{r_S^2}{r_{SE}^2} \qquad (2.8)$$

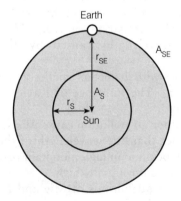

Figure 2.2 *The Radiant Power through the Surface of a Sphere with Radius r_{SE} is the Same as through the Surface of the Sun*

This determines the extraterrestrial irradiance experienced at Earth's orbital distance from the sun. However, the distance between the sun and Earth is not constant throughout the year. It varies between $1.47 \cdot 10^8$ km and $1.52 \cdot 10^8$ km. This causes a variation in the irradiance, E_e, of between 1325 W/m^2 and 1420 W/m^2. The average value, called the *solar constant*, E_0, is:

$$E_0 = 1.367 \pm 2 \text{ W/m}^2 \tag{2.9}$$

This value can be measured outside the Earth's atmosphere on a surface perpendicular to the solar radiation. The symbol I_0 is also used for the solar constant.

SOLAR IRRADIANCE ON THE SURFACE OF THE EARTH

Values measured on the surface of Earth are usually lower than the solar constant. Various influences of the atmosphere reduce the irradiance. They are:

- reduction due to reflection by the atmosphere
- reduction due to absorption in the atmosphere (mainly O_3, H_2O, O_2 and CO_2)
- reduction due to Rayleigh scattering
- reduction due to Mie scattering.

The *absorption* of light by different gases in the atmosphere, such as water vapour, ozone and carbon dioxide, is highly selective and influences only some parts of the spectrum. Figure 2.3 shows the spectrum outside the atmosphere (*AM* 0) and at the surface of the Earth (*AM* 1.5). The spectrum describes the composition of the light and the contribution of the different wavelengths to the total irradiance. Seven per cent of the extraterrestrial spectrum (*AM* 0) falls in the ultraviolet range, 47 per cent in the visible range and 46 per cent in the infrared range. The terrestrial spectrum *AM* 1.5 shows significant reductions at certain wavelengths caused by absorption by different atmospheric gases.

Molecular air particles with diameters smaller than the wavelength of light cause *Rayleigh scattering*. The influence of Rayleigh scattering rises with decreasing light wavelength.

Dust particles and other air pollution cause *Mie scattering*. The diameter of these particles is larger than the wavelength of the light. Mie scattering depends significantly on location; in high mountain regions it is relatively low, whereas in industrial regions it is usually high.

Table 2.4 shows the contributions of Mie and Rayleigh scattering and absorption for different sun heights γ_S (see section on calculating the position of the sun, p55). Climatic influences such as clouds, snow, rain or fog can cause additional reductions.

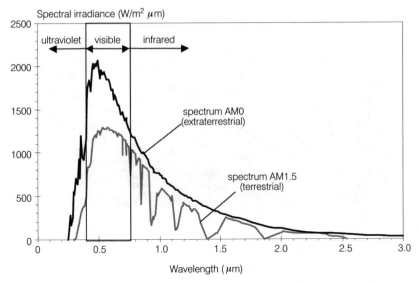

Note: AM0 is the extraterrestrial spectrum; AM1.5 is the spectrum on the Earth's surface at a sun height of 41.8°

Figure 2.3 *Spectrum of Sunlight*

The relationship between the sun height γ_S and the air mass (AM) is:

$$AM = \frac{1}{\sin \gamma_S} \qquad (2.10)$$

The AM value is a unitless measure of the length of the path of light through the atmosphere; it is expressed in multiples of the thickness of atmosphere. If the sun is at its zenith, AM is equal to 1, i.e. the light is passing vertically through the atmosphere. The AM value outside the atmosphere is zero. Figure 2.4 shows the highest position of the sun at solar noon and the corresponding AM values for various days of a year for Berlin and Cairo.

The elevation of the sun also influences the irradiation received at the surface of the Earth, which is thus dependent on the time of the year. Clouds

Table 2.4 *Reduction Influences at Different Sun Heights*

Sun Height (γ_S)	Air Mass (AM)	Absorption (%)	Rayleigh scattering (%)	Mie scattering (%)	Total reduction (%)
90°	1.00	8.7	9.4	0–25.6	17.3–38.5
60°	1.15	9.2	10.5	0.7–29.5	19.4–42.8
30°	2.00	11.2	16.3	4.1–44.9	28.8–59.1
10°	5.76	16.2	31.9	15.4–74.3	51.8–85.4
5°	11.5	19.5	42.5	24.6–86.5	65.1–93.8

Source: according to Schulze, 1970

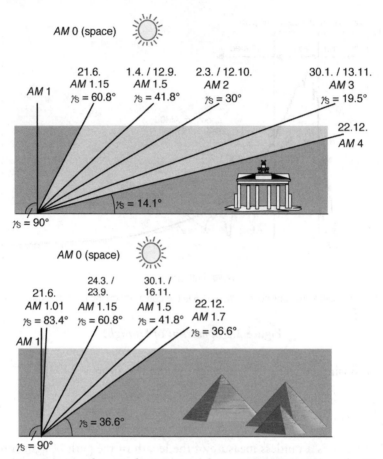

Figure 2.4 *Sun Height at Solar Noon and Air Mass (AM) Values for Various Dates in Berlin (top) and Cairo (bottom)*

and weather are important as well. The daily irradiation in central Europe can reach values above 7.5 kWh/(m² day) in summer, whereas single days in winter can have less than 0.1 kWh/(m² day). Figure 2.5 shows the variation in the irradiance for a cloudless day in summer (2 July) and in winter (28 December) as well as a very cloudy day in winter (22 December) for Karlsruhe in southern Germany.

The annual irradiation varies significantly throughout the world. For instance, in Europe there are large differences between north and south. In the north, differences between summer and winter are much higher than in the south. In Bergen (Norway, 60.4°N) the ratio of global irradiation (total irradiance on a horizontal surface on Earth) in June to global irradiation in December is 40:1, whereas in Lisbon (Portugal, 38.72°N) this ratio is only 3.3:1. Central and northern Europe have annual global irradiation values of between 700 kWh/(m² year) and 1000 kWh/(m² year). In southern Europe this irradiation can be more than 1700 kWh/(m² year) and in desert regions of

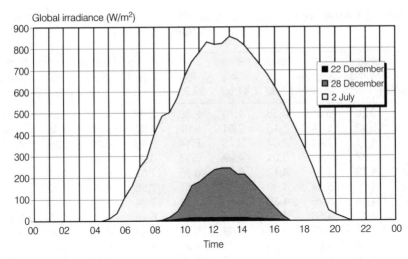

Figure 2.5 *Global Irradiance throughout the Day in Karlsruhe (Germany) for 2 July and 22 and 28 December 1991*

the Earth's sunbelt the figure is around 2500 kWh/(m² year). However, the latitude can only give a rough indication of the annual irradiation because local effects have a major impact on the energy reaching Earth's surface. For instance, the annual irradiation in Stockholm (Sweden) and Berlin (Germany) are nearly the same, although Stockholm's latitude is 7° higher than Berlin. On the other hand, the annual irradiation in London, which is south of Berlin, is significantly lower.

Table 2.5 gives an overview of monthly average global irradiation values for some locations around the world. It clearly demonstrates that there are significant variations between different locations. However, the precise irradiance at the given site is required for planning solar energy systems. This can be estimated using existing databases. Some free Internet databases offer monthly irradiation values for many locations in the world; some even offer hourly irradiance datasets for some sites (e.g. *www.satellight.com*, *eosweb.larc.nasa.gov/sse* or *rredc.nrel.gov/solar*). Computer programs (see also CD-ROM) such as the Meteonorm program can also be used for interpolating the meteorological parameters of a given site based on measurements taken at locations close to the proposed site.

The annual irradiation in the Sahara is about 2350 kWh/(m² year). The total annual irradiation received by the surface of the Sahara (around 8.7 million km²) is nearly 200 times higher than the global annual primary energy demand; in fact the global primary energy demand could be provided by collecting the solar energy received by 48,500 km² of the Sahara, an area slightly larger than Switzerland, or one-ninth that of California. These numbers clearly show that it is possible to provide the whole global energy demand solely by solar energy.

Table 2.5 *Monthly Average Values in kWh/(m² day) of the Daily Global Irradiation*

Latitude	Bergen Norway 60.40°N	Berlin Germany 52.47°N	London UK 51.52°N	Rome Italy 41.80°N	LA US 33.93°N	Cairo Egypt 30.08°N	Bombay India 19.12°N	Upington RSA 28.40°S	Sydney Australia 33.95°S
Jan	0.20	0.61	0.56	1.70	2.88	3.09	4.74	8.08	6.41
Feb	0.72	1.14	1.10	2.54	3.97	4.00	5.56	7.45	5.57
Mar	1.71	2.44	2.07	3.78	5.14	5.15	6.29	6.26	4.72
Apr	3.27	3.49	3.04	4.99	6.47	6.27	6.72	5.19	3.47
May	4.13	4.77	4.12	6.03	6.55	7.03	6.77	4.26	2.63
June	4.85	5.44	4.99	6.59	6.57	7.56	4.99	3.72	2.38
July	4.15	5.26	4.38	6.86	7.38	7.34	3.84	4.04	2.52
Aug	3.49	4.58	3.62	6.16	6.82	6.76	3.86	4.95	3.47
Sep	1.86	3.05	2.71	4.69	5.26	5.87	4.65	6.09	4.66
Oct	0.94	1.59	1.56	3.29	4.24	4.69	5.11	7.21	5.63
Nov	0.30	0.76	0.81	2.02	3.22	3.45	4.73	8.27	6.40
Dec	0.12	0.45	0.47	1.51	2.72	2.86	4.46	8.49	6.69
Average	2.15	2.81	2.46	4.19	5.10	5.34	5.14	6.17	4.55

Source: data from Palz and Greif, 1996; NASA, 2003

IRRADIANCE ON A HORIZONTAL PLANE

As described above, solar radiation is scattered and reflected when passing through the atmosphere. Rays of extraterrestrial sunlight are virtually parallel. Terrestrial sunlight, on the other hand, consists of direct and diffuse components (see Figure 2.6). Direct solar radiation casts shadows, because it is directional, coming directly from the sun; diffuse irradiation, on the other hand, has no defined direction. The total irradiance on a horizontal surface on Earth is also called global irradiance $E_{G,hor}$. It is the sum of the direct irradiance $E_{dir,hor}$ and the diffuse irradiance $E_{diff,hor}$ on the horizontal surface:

$$E_{G,hor} = E_{dir,hor} + E_{diff,hor} \qquad (2.11)$$

Table 2.6 shows the monthly average daily direct and diffuse irradiation in Berlin and Cairo. In Berlin the diffuse irradiation dominates, whereas the direct irradiation is much higher in Cairo, even in winter.

The annual diffuse irradiation does not have to vary greatly between locations despite high differences in the annual global irradiation (see Table 2.7). Upington in South Africa and London in the UK have annually about the same amount of diffuse irradiation even though the annual global irradiation in Upington is more than twice that of London. Regions with high air pollution or tropical regions have a significantly elevated contribution of diffuse irradiation. However, differences in annual direct irradiations are much higher. For example, the annual direct irradiation in Upington is almost five times that of London.

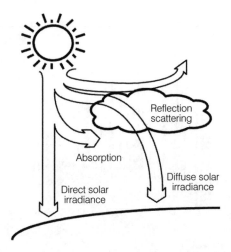

Figure 2.6 *Sunlight Passing Through the Atmosphere*

Days with low global irradiation have a high percentage of diffuse irradiation, sometimes approaching 100 per cent; however, the contribution of diffuse irradiation decreases to less than 20 per cent on days with very high global irradiation values. Figure 2.7 shows the variation of direct and diffuse irradiation throughout one year in Berlin as an example of a location with relatively low annual irradiation. Figure 2.8 shows irradiation values for Cairo as an example for a location with relatively high annual irradiation. There are significant differences between these locations. In Berlin, variation of both diffuse and direct irradiation is much higher throughout the year than in Cairo.

Table 2.6 *Monthly Average Daily Direct and Diffuse Irradiation in $kWh/(m^2\ day)$ in Berlin and Cairo*

		Jan	Feb	Mar	Apr	May	June	July	Aug	Sep	Oct	Nov	Dec	Average
Berlin	Direct	0.17	0.40	1.03	1.42	2.13	2.58	2.29	2.05	1.38	0.54	0.22	0.10	1.20
	Diffuse	0.44	0.74	1.41	2.07	2.64	2.86	2.97	2.53	1.67	1.05	0.54	0.35	1.61
Cairo	Direct	1.74	2.37	3.07	3.78	4.56	5.16	4.93	4.57	3.86	3.07	1.96	1.58	3.39
	Diffuse	1.35	1.63	2.08	2.49	2.47	2.40	2.41	2.19	2.01	1.62	1.49	1.28	1.95

Table 2.7 *Annual Average Daily Direct and Diffuse Irradiation in $kWh/(m^2\ day)$*

	Bergen	Berlin	London	Rome	LA	Cairo	Bombay	Upington	Sydney
Direct	0.86	1.20	0.99	2.41	3.03	3.39	2.75	4.70	2.42
Diffuse	1.29	1.61	1.47	1.78	2.07	1.95	2.39	1.47	2.13

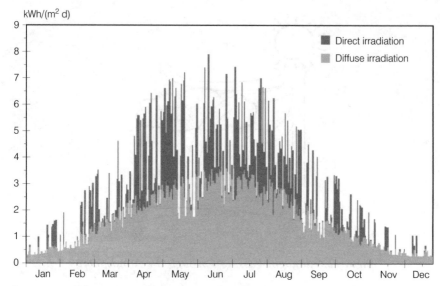

Figure 2.7 *Daily Direct and Diffuse Irradiation in Berlin*

Many meteorological stations measure global irradiance only. However, most calculations for solar energy systems need a separation into direct and diffuse irradiance. Empirical functions found by statistical investigations can split the global irradiation into direct and diffuse irradiation (Reindl et al, 1989). Hourly values of the global irradiance $E_{G,hor}$, the extraterrestrial irradiance E_0 and the sun height γ_S (which is calculated in the following section) define the factor k_T as follows:

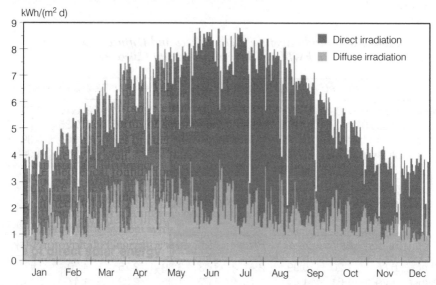

Figure 2.8 *Daily Direct and Diffuse Irradiation in Cairo*

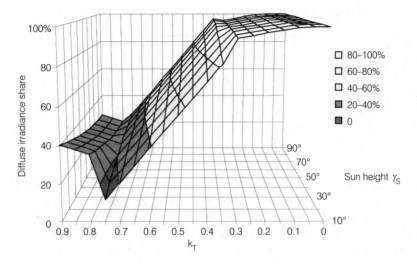

Figure 2.9 *Diffuse Irradiance Component as a Function of* k_T *and* γ_S

$$k_T = \frac{E_{G,hor}}{E_0 \cdot \sin \gamma_S} \qquad (2.12)$$

With this factor, the diffuse irradiance $E_{diff,hor}$ can be calculated easily using the global irradiance $E_{G,hor}$ and the sun height γ_S:

$$E_{diff,hor} = E_{G,hor} \cdot (1.020 - 0.254 \cdot k_T + 0.0123 \cdot \sin \gamma_S) \qquad \text{for } k_T \leq 0.3$$

$$E_{diff,hor} = E_{G,hor} \cdot (1.400 - 1.749 \cdot k_T + 0.177 \cdot \sin \gamma_S) \qquad \text{for } 0.3 < k_T < 0.78$$

$$E_{diff,hor} = E_{G,hor} \cdot (0.486 \cdot k_T - 0.182 \cdot \sin \gamma_S) \qquad \text{for } k_T \geq 0.78.$$
$$(2.13)$$

Figure 2.9 shows this correlation graphically. It is obvious that the diffuse irradiation component is very low if the global irradiance values are high on clear days ($k_T \rightarrow 1$); however, the diffuse irradiation component is rarely below 20 per cent. If it is very cloudy and the global irradiance is low ($k_T \rightarrow 0$), the diffuse irradiance component can reach 100 per cent. The following section describes methods to calculate the solar altitude angle, or sun height γ_S.

CALCULATION OF THE SUN'S POSITION

The position of the sun is essential for many further calculations for solar energy systems. The two angles *sun height* (solar altitude or elevation) γ_S and solar or *sun azimuth* α_S define the position of the sun. However, definitions for these angles and the symbols used vary in the literature. The convention used

Table 2.8 *Different Definitions of Solar Azimuth Angle*

Reference	Symbol	N	NE	E	SE	S	SW	W	NW
ISES, 2001; NREL, 2000	γ	0°	45°	90°	135°	180°	225°	270°	315°
CEN, 1999 for $\varphi>0$[a]	γ	180°	225°	270°	315°	0°	45°	90°	135°
CEN, 1999 for $\varphi<0$[b]	γ	0°	270°	270°	225°	180°	135°	90°	45°
DIN, 1985, this book	α_S	0°	45°	90°	135°	180°	225°	270°	315°

Note: φ, latitude; a north of the equator; b south of the equator

in this book defines the sun height as the angle between the centre of the sun and the horizon seen by an observer. The azimuth angle of the sun describes the angle between geographical north and the vertical circle through the centre of the sun. EN ISO 9488 defines the solar azimuth as the angle between the apparent position of the sun and south measured clockwise at the northern hemisphere and between the apparent position of the sun and north measured anticlockwise in the southern hemisphere (CEN, 1999). This can cause confusion when comparing calculations with different angle definitions. Table 2.8 shows the different solar azimuth angle definitions of various sources.

The sun height and solar azimuth depend on the geographical location of the observer, the date, time and time zone. The position of the sun is strongly influenced by the angle between the equatorial plane of the Earth and the rotational plane of the Earth around the sun, called the solar declination. The solar declination δ varies between +23°26.5′ and –23°26.5′ over a year. Since the orbit of the earth around the sun is not circular, the length of a solar day also changes throughout the year. Usually, the so-called equation of time *eqt* takes this into consideration. Many algorithms have been developed to calculate the position of the sun. A relatively simple algorithm is described overleaf (DIN, 1985).

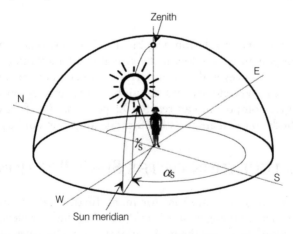

Figure 2.10 *Definitions of the Angles Describing the Position of the Sun Used in this Book*

With the day angle

$$y' = 360° \cdot \frac{\text{day of the year}}{\text{number of days of the year}}$$

the solar declination becomes:

$$\delta(y') = \{0.3948 - 23.2559 \cdot \cos(y'+9.1°) - 0.3915 \cdot \cos(2 \times y'+5.4°) - 0.1764 \cdot \cos(3 \cdot y'+105.2°)\}°$$

$$(2.14)$$

and the equation of time

$$eqt(y') = [0.0066 + 7.3525 \cdot \cos(y'+85.9°) + 9.9359 \cdot \cos(2 \cdot y'+108.9°) + 0.3387 \cdot \cos(3 \cdot y'+105.2°)]\text{min}$$

$$(2.15)$$

is calculated. With the *Local time*, the *Time zone* (e.g. Greenwich mean time GMT = 0, Central European Time CET = 1 h, Pacific Standard Time PST = −8 h) and the longitude λ, the mean local time MLT becomes:

$$\text{MLT} = \text{Local time} - \text{Time zone} + 4 \cdot \lambda \cdot \text{min/°} \qquad (2.16)$$

Adding the equation of time *eqt* to the mean local time *MLT* provides the *Solar time*:

$$\text{Solar time} = \text{MLT} + eqt \qquad (2.17)$$

With the latitude φ of the location and the hour angle ω:

$$\omega = (12.00 \text{ h} - \text{Solar time}) \cdot 15°/\text{h} \qquad (2.18)$$

the angle of solar altitude (sun height) γ_S and angle of solar azimuth α_S become:

$$\gamma_S = \arcsin(\cos\omega \cdot \cos\varphi \cdot \cos\delta + \sin\varphi \cdot \sin\delta) \qquad (2.19)$$

$$\alpha_S = \begin{cases} 180° - \arccos \dfrac{\sin\gamma_S \cdot \sin\varphi_S - \sin\delta}{\cos\gamma_S \cdot \cos\varphi} & \text{if Solar time} \leq 12.00 \text{ h} \\[2ex] 180° + \arccos \dfrac{\sin\gamma_S \cdot \sin\varphi_S - \sin\delta}{\cos\gamma_S \cdot \cos\varphi} & \text{if Solar time} > 12.00 \text{ h} \end{cases}$$

$$(2.20)$$

Table 2.9 *Latitude φ and Longitude λ of Selected Locations*

	Bergen	Berlin	London	Rome	LA	Cairo	Bombay	Upington	Sydney
φ	60.40°	52.47°	51.52°	41.80°	33.93°	30.08°	19.12°	−28.40°	−33.95
λ	5.32°	13.30°	−0.11°	12.58°	−118.40°	31.28°	72.85°	21.27°	151.18°

Table 2.9 shows angles of latitude and longitude for selected locations.

Other algorithms such as the SUNAE algorithm (Walraven, 1978; Wilkinson, 1981; Kambezidis and Papanikolaou, 1990) or the NREL SOLPOS algorithm (NREL, 2000) have improved accuracies, mainly at low solar altitudes. These algorithms include the refraction of the beam irradiance by the atmosphere. However, they are also much more complex than the algorithm described above. The CD-ROM of the book contains the code for these algorithms.

Solar position or *sun-path diagrams* are used to visualize the path of the sun in the course of a day. These diagrams show sun height and azimuth for every hour of the selected days with a curve drawn through the points. Figure 2.11 shows the solar position diagram for Berlin and Figure 2.12 that for Cairo. For clarity, these diagrams show only five months from the first half of the year; the corresponding months for the second half of the year have nearly symmetrical curves.

Solar position diagrams for southern latitudes look similar, except that the south position is in the centre of the diagram instead of north. Solar position diagrams for locations between the northern and southern tropics are different in that the sun is in the south at solar noon for some months and in the north for others.

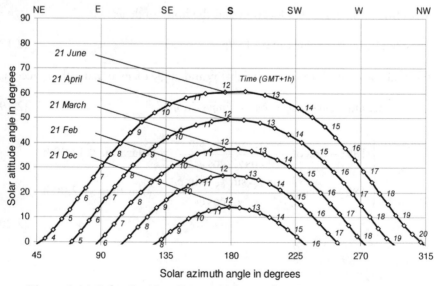

Figure 2.11 *Solar Position Diagram for Berlin, Germany (52.5°N)*

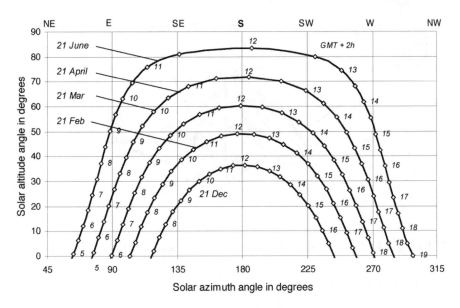

Figure 2.12 *Solar Position Diagram for Cairo, Egypt (30.1°N)*

CALCULATION OF THE SOLAR ANGLE OF INCIDENCE

The solar angle of incidence θ_{hor} on a horizontal surface is a direct function of the sun height γ_S. This angle is also called the *zenith angle* θ_Z:

$$\theta_{hor} = \theta_Z = 90° - \gamma_S \tag{2.21}$$

The calculation of the angle of incidence θ_{tilt} on a tilted surface is more complicated. The surface azimuth angle α_t describes the deviation from the south. If the surface faces to the west, α_t is positive. The inclination angle γ_t describes the surface tilt or slope of the surface. If the surface is horizontal, γ_t is zero. Figure 2.13 visualizes these angles.

The angle of incidence θ_{tilt} is the angle between the vector s in the direction of the sun and the normal vector n perpendicular to the surface. The position of the sun has been defined in spherical coordinates and thus must be transformed into Cartesian coordinates with the base vectors north, west and zenith for further calculations. The vectors s and n become:

$$s = (\cos\alpha_S \cdot \cos\gamma_S, - \sin\alpha_S \cdot \cos\gamma_S, \sin\gamma_S)^T \tag{2.22}$$

$$n = (-\cos\alpha_t \cdot \sin\gamma_t, \sin\alpha_t \cdot \sin\gamma_t, \cos\gamma_t)^T \tag{2.23}$$

Both vectors are normalized, and thus the solar angle of incidence θ_{tilt} on a tilted surface is obtained by calculating the scalar multiplication of these two vectors:

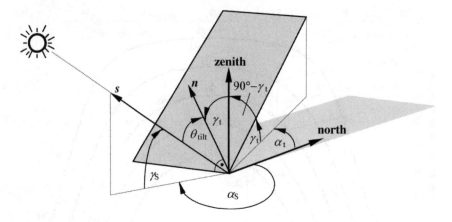

Figure 2.13 *Definition of the Solar Angle of Incidence on a Tilted Surface*

$\theta_{tilt} = \arccos(s \cdot n) =$

$= \arccos(-\cos\alpha_S \cdot \cos\gamma_S \cdot \cos\alpha_t \cdot \sin\gamma_t -$

$\sin\alpha_S \cdot \cos\gamma_S \cdot \sin\alpha_t \cdot \sin\gamma_t + \sin\gamma_S \cdot \cos\gamma_t)$

$= \arccos) - \cos\gamma_S \cdot \sin\gamma_t \cdot \cos(\alpha_S - \alpha_t) + \sin\gamma_S \cdot \cos\gamma_t)$ (2.24)

IRRADIANCE ON TILTED SURFACES

The global irradiance $E_{G,tilt}$ on a tilted surface is composed of the direct irradiance $E_{dir,tilt}$, the diffuse sky irradiance $E_{diff,tilt}$ and the ground reflection $E_{refl,tilt}$ (a factor that does not exist for horizontal surfaces):

$$E_{G,tilt} = E_{dir,tilt} + E_{diff,tilt} + E_{refl,tilt}$$ (2.25)

Direct irradiance on tilted surfaces

The horizontal surface in Figure 2.14 with the area A_{hor} receives the same radiant power Φ as a smaller area A_s which is normal (perpendicular) to the incoming sunlight.

With

$$\Phi_{dir,hor} = E_{dir,hor} \cdot A_{hor} = \Phi_{dir,s} = E_{dir,s} \cdot A_s$$

and

$$A_s = A_{hor} \cdot \cos\theta_{hor} = A_{hor} \cdot \sin\gamma_S$$

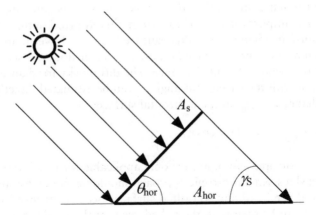

Figure 2.14 *Irradiance on a Horizontal Area A_{hor} and an Area A_s Perpendicular to the Sunlight*

it follows that:

$$E_{dir,s} = \frac{E_{G,hor}}{\sin \gamma_S} \geqslant E_{dir,hor} \tag{2.26}$$

It becomes apparent that the direct normal or *beam irradiance* $E_{dir,s}$ on a surface perpendicular to the path of the light is higher than the direct irradiance $E_{dir,hor}$ on a horizontal surface; this fact is taken into account when planning solar energy systems. Inclining the surface of the system increases the energy yield, especially at high latitudes with low solar height angles.

With θ_{tilt} from (2.24), the direct irradiance on a tilted surface is $E_{dir,s} = \dfrac{E_{dir,tilt}}{\cos \theta_{tilt}}$.

The direct irradiation of a tilted surface can be calculated directly from the direct irradiation on the horizontal surface:

$$E_{dir,tilt} = E_{dir,hor} \cdot \frac{\cos \theta_{tilt}}{\sin \gamma_S} \tag{2.27}$$

However, for low sun heights, small variations of the horizontal irradiance can cause unrealistically high irradiances on tilted surfaces. Therefore, it should always be checked that the calculated direct irradiance on a tilted plane is below a maximum threshold.

Diffuse irradiance on tilted surfaces

There are two approaches for estimating the diffuse sky irradiance $E_{diff,tilt}$ on a tilted surface: the isotropic approach and the anisotropic approach. The isotropic approach assumes that the diffuse radiance is constant across the whole sky. An important conclusion from this approximation is that the isotropic diffuse irradiance on a tilted surface is always lower than the diffuse irradiance on a horizontal surface because a receiver does not see the diffuse irradiance from behind the tilted surface. The diffuse sky irradiance $E_{diff,tilt}$ on a tilted surface with the surface tilt angle γ_t can be estimated directly from the diffuse irradiance $E_{diff,hor}$ on the horizontal surface:

$$E_{dir,tilt} = E_{dir,hor} \cdot \frac{1}{2} \cdot (1 + \cos\gamma_t) \tag{2.28}$$

However, the isotropic assumption is only applicable for rough estimations or very overcast skies. An anisotropic approach should be chosen for more precise calculations of the irradiance on tilted surfaces because diffuse irradiance is directional. It can be seen with the naked eye that the brightness increases at the horizon and near the sun. Two models, which consider these effects, are described here.

Klucher's model (Klucher, 1979) calculates the diffuse irradiance $E_{diff,tilt}$ on a tilted surface in a relatively simple way:

With $F = 1 - \left(\dfrac{E_{dir,hor}}{E_{G,hor}}\right)^2$, the diffuse irradiance is given by:

$$E_{dir,tilt} = E_{dir,hor} \cdot \frac{1}{2} \cdot (1 + \cos\gamma_t) \cdot \left(1 + F + \sin^3\right) \frac{\gamma_t}{2} \cdot$$

$$(1 + F \cdot \cos^2\theta_{tilt} \cdot \cos^3\gamma_S) \tag{2.29}$$

This model gives a relatively good estimation of the diffuse irradiance. A more precise model for the calculation of the diffuse irradiation on a tilted surface is the so-called *Perez model* (Perez and Stewart, 1986; Perez et al, 1987; Perez et al, 1990). However, this model is also more complex.

The Perez model defines an atmospheric clearness index ε and an atmospheric brightness parameter Δ that uses the angle of incidence θ_{hor} of the sunlight on the horizontal surface (measured in rad), the constant $\kappa = 1.041$, the solar constant E_0, the direct and diffuse irradiance on the horizontal surface as well as the air mass Air Mass ($AM = 1/\sin\gamma_S$):

$$\varepsilon = \dfrac{\dfrac{E_{diff,hor} + E_{dir,hor} \cdot \sin^{-1}\gamma_t}{E_{dir,hor}} + \kappa \cdot \theta^3_{hor}}{1 + \kappa \cdot \theta^3_{hor}} \tag{2.30}$$

Table 2.10 *Constants for Estimating* F_1 *and* F_2 *as a Function of* ε

ε class	1	2	3	4	5	6	7	8
ε	1.000– 1.065	1.065– 1.230	1.230– 1.500	1.500– 1.950	1.950– 2.800	2.800– 4.500	4.500– 6.200	6.200– ∞
F_{11}	–0.008	0.130	0.330	0.568	0.873	1.132	1.060	0.678
F_{12}	0.588	0.683	0.487	0.187	–0.392	–1.237	–1.600	–0.327
F_{13}	–0.062	–0.151	–0.221	–0.295	–0.362	–0.412	–0.359	–0.250
F_{21}	–0.060	–0.019	0.055	0.109	0.226	0.288	0.264	0.156
F_{22}	0.072	0.066	–0.064	–0.152	–0.462	–0.823	–1.127	–1.377
F_{23}	–0.022	–0.029	–0.026	–0.014	0.001	0.056	0.131	0.251

Source: Perez et al, 1990

$$\Delta = AM \cdot \frac{E_{\text{diff,hor}}}{E_0} \tag{2.31}$$

With these parameters, the circumsolar brightening coefficient F_1 and the horizon brightening coefficient F_2 can be calculated:

$$F_1 = F_{11}(\varepsilon) + F_{12}(\varepsilon) \cdot \Delta + F_{13}(\varepsilon) \cdot \theta_{\text{hor}} \tag{2.32}$$

$$F_2 = F_{21}(\varepsilon) + F_{22}(\varepsilon) \cdot \Delta + F_{23}(\varepsilon) \cdot \theta_{\text{hor}} \tag{2.33}$$

The constants F_{11} to F_{23} are estimated from Table 2.10. They vary according to the eight different atmospheric clearness classes (ε class = 1–8) that correspond to the equivalent atmospheric clearness index values ε.

With F_1 and F_2 as well as a = max (0; cos θ_{gen}) and b = max (0.087; sinγ_S), the diffuse irradiance $E_{\text{diff,tilt}}$ on a tilted plane using the diffuse irradiance $E_{\text{diff,hor}}$ on a horizontal plane becomes:

$$E_{\text{diff,tilt}} = E_{\text{dir,hor}} \cdot \left[\frac{1}{2} \cdot (1 + \cos \gamma_t) \cdot (1 - F_1) + \frac{a}{b} \cdot F_1 + F_2 \cdot \sin \gamma_t \right] \tag{2.34}$$

Ground reflection

For calculating the ground reflection $E_{\text{refl,tilt}}$ an isotropic approach is sufficient. Anisotropic approaches have shown only insignificant improvements. With the global irradiance $E_{\text{G,hor}}$ on a horizontal surface and the albedo A, the ground reflected irradiance $E_{\text{refl,tilt}}$ on a surface with tilt angle γ_t becomes:

$$E_{\text{refl,tilt}} = E_{\text{G,hor}} \cdot A \cdot \frac{1}{2} (1 - \cos \gamma_t) \tag{2.35}$$

Table 2.11 *Albedo for Different Types of Surface*

Surface	Albedo A	Surface	Albedo A
Grass (Summer)	0.25	Asphalt	0.15
Lawn	0.18–0.23	Woods	0.05–0.18
Dry grass	0.28–0.32	Heathland and sand	0.10–0.25
Uncultivated fields	0.26	Water surface ($\gamma_S > 45°$)	0.05
Soil	0.17	Water surface ($\gamma_S > 30°$)	0.08
Gravel	0.18	Water surface ($\gamma_S > 20°$)	0.12
Concrete, weathered	0.20	Water surface ($\gamma_S > 10°$)	0.22
Concrete, clean	0.30	Fresh snow cover	0.80–0.90
Cement, clean	0.55	Old snow cover	0.45–0.70

Source: Dietze, 1957, TÜV 1984

The albedo value A influences the accuracy of the calculations noticeably. Table 2.11 shows approximate albedo values for different types of surface. For unknown surfaces, the value $A = 0.2$ is often used.

Irradiance gain for surface tilt or tracking

If a solar energy system tracks the sun so that the angle of incidence is virtually zero, the energy yield increases significantly. The higher direct irradiance on a surface perpendicular to the solar radiation causes the increased energy yield. During days with high direct irradiation, tracking can achieve energy gains over horizontal orientation in the order of 50 per cent in summer and up to 300 per cent in winter, depending on the latitude of the location (see Figure 2.15). However, tracking can cause a reduction in energy yield in overcast conditions because the contribution of diffuse irradiation from behind the surface is lost. Tracking achieves the main energy gain in summer. On the one hand, the absolute energy gain in summer is higher than in winter; on the other hand the number of overcast days is usually lower in summer.

There are two principle options for solar energy system tracking: one-axis and two-axis tracking. Two-axis tracking systems move a surface always into an ideal position; however, two-axis tracking systems are relatively complicated and thus one-axis tracking systems are preferred in some instances. One-axis tracking systems can follow the daily or annual path of the sun. Tracking to the annual path of the sun is relatively simple: the surface tilt angle need only be changed once a week or even once a month.

A *two-axis tracked solar energy system* installed in middle-European latitudes can obtain an energy gain in the order of 30 per cent compared to inclined non-tracking systems. The energy gain of a one-axis tracking system is slightly lower; it is closer to 20 per cent. Regions with higher annual irradiation also have a higher absolute energy gain because typically the direct beam contribution is much higher. However, tracking systems are more complicated, more expensive and have higher operating and maintenance costs. The tracking system must also resist strong winds. There are two

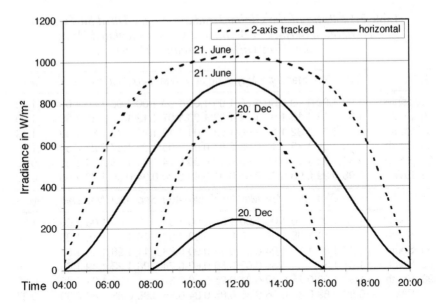

Figure 2.15 *Irradiance on Horizontal and Two-axis Tracked Surfaces for Cloudless Days at a Site at 50° Latitude*

methods for achieving the motion required for tracking: electrical motors and thermohydraulic systems. The electric motor driving the tracking unit needs electrical energy and thus reduces the energy gain of the system. Prototypes with thermohydraulic drives have shown some operational problems. If the tracking system fails, more often than not it is stuck in an inefficient position and hence the system output is very poor until the system is repaired.

The energy gain of tracked solar energy systems does not usually compensate for the disadvantages. Hence, there are only few operational tracking systems at present. Only large systems in regions with a very high annual irradiation can achieve economical advantages from tracking (Quaschning and Ortmanns, 2003).

The situation is totally different for concentrating solar energy systems in which optical systems concentrate the energy to a much smaller area. Such systems are in operation for solar thermal troughs, solar tower power plants (see section headed 'Use of direct solar energy') or concentrating photovoltaic systems. These systems have very narrow angles of acceptance for incoming irradiance and thus do not work satisfactorily without tracking. However, most concentrating systems can only use direct solar irradiation.

Using the optimal orientation for non-tracked solar energy systems can also increase the energy gain significantly. The optimal orientation for solar energy systems operating the whole year at latitudes higher than 30° is about 30° to the south in the northern hemisphere and 30° to the north in the southern hemisphere. The optimum tilt angles for systems that only work in summer are flatter, but are much steeper for those that only work in winter.

Table 2.12 *Ratio of the Global Irradiation on a Tilted Surface to a Horizontal Surface in Berlin and Cairo Calculated Using the Perez Diffuse Irradiance Model*

Berlin	Jan	Feb	Mar	Apr	May	June	July	Aug	Sep	Oct	Nov	Dec	Average
Horizontal	1.00	1.00	1.00	1.00	1.00	1.00	1.00	1.00	1.00	1.00	1.00	1.00	1.00
10° South	1.33	1.23	1.13	1.08	1.04	1.03	1.03	1.07	1.12	1.17	1.30	1.36	1.09
30° South	1.90	1.61	1.32	1.17	1.06	1.03	1.04	1.13	1.28	1.43	1.81	1.99	1.19
60° South	2.39	1.88	1.38	1.11	0.93	0.87	0.90	1.03	1.29	1.57	2.23	2.54	1.15
90° South	2.34	1.73	1.15	0.83	0.63	0.58	0.60	0.74	1.04	1.38	2.14	2.51	0.89
45° E/W	0.98	0.99	0.98	0.92	0.91	0.91	0.90	0.90	0.94	0.95	1.02	1.03	0.93

Cairo	Jan	Feb	Mar	Apr	May	June	July	Aug	Sep	Oct	Nov	Dec	Average
Horizontal	1.00	1.00	1.00	1.00	1.00	1.00	1.00	1.00	1.00	1.00	1.00	1.00	1.00
10° South	1.19	1.14	1.08	1.04	1.00	0.98	0.99	1.02	1.06	1.12	1.16	1.20	1.06
30° South	1.47	1.33	1.17	1.04	0.93	0.88	0.90	0.98	1.11	1.26	1.40	1.51	1.10
60° South	1.61	1.36	1.08	0.85	0.67	0.59	0.62	0.74	0.96	1.23	1.49	1.68	0.96
90° South	1.37	1.08	0.75	0.49	0.31	0.25	0.26	0.37	0.61	0.93	1.24	1.45	0.63
45° E/W	0.89	0.89	0.88	0.86	0.86	0.85	0.84	0.85	0.86	0.88	0.88	0.90	0.87

However, near the equator an almost horizontal plane receives the highest annual irradiation. Table 2.12 shows the energy production over the year for some surface orientations in Berlin and Cairo.

Solar energy systems are often installed on pitched roofs, which are usually not at the optimum tilt angle. A badly oriented roof can reduce the energy production significantly; however, the acceptable orientations for the roof have a rather large tolerance, as shown in Figure 2.16 and Figure 2.17 for Berlin and Cairo.

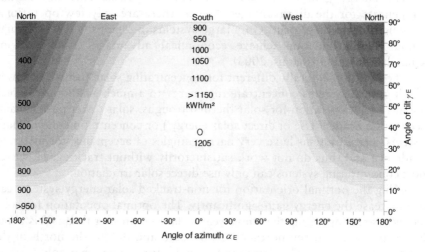

Figure 2.16 *Annual Irradiation on Various Inclined Surfaces in Berlin (52.5°N)*

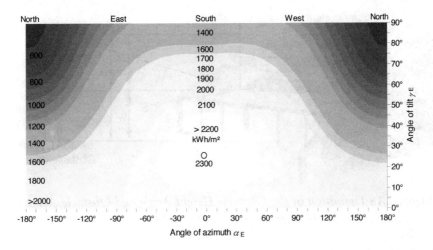

Figure 2.17 *Annual Irradiation on Various Inclined Surfaces in Cairo (30.1°N)*

In the southern hemisphere, the optimum tilt angles are similar, with the optimum surface azimuth facing the equator, i.e. north. Close to the equator, very flat tilt angles are ideal because the sun is in the zenith for long periods of time. Therefore, the irradiation losses for vertical surfaces are much higher at low latitudes.

CALCULATION OF SHADING LOSSES

All calculations in the previous sections have assumed that the irradiance is not reduced as a result of shadows of objects in the surroundings. Totally unshaded surfaces are difficult to find in reality. Shading can cause high losses of performance for solar energy systems, in particular for photovoltaic systems. Therefore, the following sections explain how shadowing can be taken into account for the irradiance calculations.

Recording the surroundings

This section uses the example of a photovoltaic system to be installed on the roof of a detached house. Various trees and bushes as well as another house are in the vicinity of the building and might affect the irradiance that could be collected by the system. First, the point of reference must be defined. The point with the highest probability of being in shadow is the best choice. All azimuth and height angles will then refer to this point.

In the next step, all azimuth and height angles of the objects in the surroundings must be estimated with respect to the point of reference. A compass can be used to find the azimuth angle α. Simple geometrical calculations can estimate the height angle γ as shown for a tree in Figure 2.18.

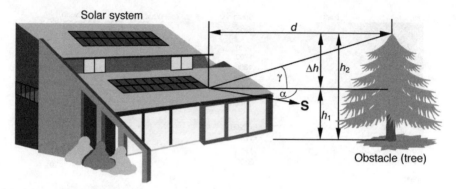

Figure 2.18 *Definition of the Obstacle Height Angle and Obstacle Azimuth Using a Freely Chosen Point of Reference*

The height angle γ is:

$$\gamma = \arctan\left(\frac{h_2 - h_1}{d}\right) = \arctan\left(\frac{\Delta h}{d}\right) \tag{2.36}$$

These calculations must be performed for all possible obstacles in the vicinity of the planned solar energy system. Specialized instruments can be used to estimate object height and azimuth angles more easily. In addition to expensive surveyor's transits, simple instruments can be also used. A solar position diagram with a trigonometrically subdivided height axis and a regular 180° azimuth axis is the basis for such an instrument. This diagram must be copied onto a transparency, which is then bent in a semicircle. The observer looks through this diagram to the objects and can directly read height and azimuth angles as shown in Figure 2.19.

It is also possible to take digital photographs of the surroundings and to use professional software to analyse it. This software automatically estimates azimuth and height angles.

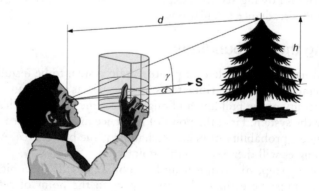

Figure 2.19 *Estimation of Object Azimuth and Height Angles Using a Simple Optical Instrument*

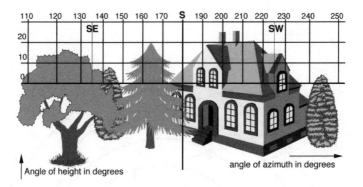

Figure 2.20 *Surroundings Seen through a Screen with Angular Grid*

Figure 2.20 shows the surroundings of the example site with a grid in spherical coordinates. A polygonal approximation of the surrounding's silhouette is drawn into the solar position diagram, as shown in Figure 2.21. It is possible to read from this diagram the times at which shading at the point of reference will occur. For instance, the observation point is shaded for almost the whole day on 21 December. Only in the morning and at noon can direct sunlight reach the observation point for about one hour. On 21 February, no shading can be expected after 9.00am. There will be no shading by surrounding objects between March and September. For clarity, the solar position diagram does not include the second half of the year.

Estimation of the degree of direct shading

For further calculations, it is assumed that the surroundings of the point of reference are given by the approximated polygonal shape shown in Figure 2.21. A horizontal ray originating at the position of the sun is used for testing whether the position of the sun is inside or outside the polygon describing the shading objects (Figure 2.22). The number of times the ray crosses the polygon boundary lines is determined. All boundary lines to the left from the position of the sun must be considered. If the number of intersections is even, the position of the sun is outside the polygon. If the number of intersections is odd, the position of the sun is inside the polygon and direct sunlight is blocked.

However, not all objects between the sun and the observer block the direct irradiance completely; in this case, the transmittance must be considered. As an example, a part of the direct irradiance may pass through trees. The transmittance τ describes the fraction that passes through trees. The transmittance of deciduous trees has been investigated (see Sattler and Sharples, 1987):

- leafless (winter) $\tau = 0.64$
- totally in leaf (summer) $\tau = 0.23$

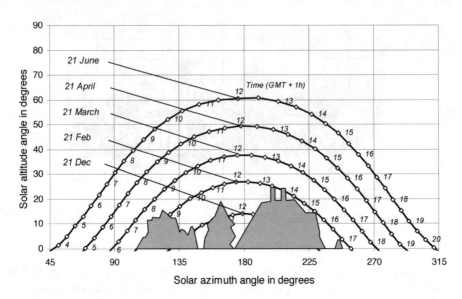

Figure 2.21 *Solar Position Diagram of Berlin with an Approximation of the Surroundings*

For further calculations, the *degree of direct shading* S_{dir} is introduced. The degree of shading describes the portion of the direct irradiance that is reduced by shading:

$$S_{dir} \begin{cases} 0 & \text{if } (\gamma_S, \alpha_S) \text{ is outside the object polygon} \\ 1 & \text{if } (\gamma_S, \alpha_S) \text{ is inside an opaque object polygon} \\ 1 - \tau & \text{if } (\gamma_S, \alpha_S) \text{ is inside a semi-transparent object polygon} \end{cases}$$

$$(2.37)$$

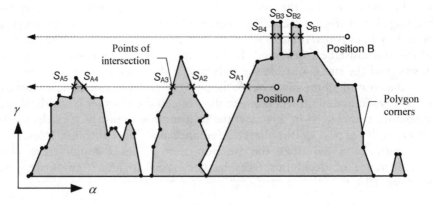

Figure 2.22 *Shading Test for Two Different Positions of the Sun A and B (Position A has 5 Intersections, Position B has 4)*

Estimation of the degree of shading of diffuse irradiance

For the estimation of the degree of shading of diffuse irradiance, the object polygon is considered as being projected on a hemisphere. The objects represented by the object polygon block the diffuse irradiation that passes through the object polygon on the hemisphere. As a simple example, Figure 2.23 shows a polygon with four points, with two of them on the horizontal ground plane. The two remaining points define the polygon clearly.

$\gamma(\alpha) = m \cdot \alpha + n$ describes the connecting line of the points p_1 and p_2 for $\alpha_1 \neq \alpha_2$ as well as $\gamma_1 > 0$ and $\gamma_2 > 0$. The parameters m and n of this line are:

$$m = \frac{\gamma_2 - \gamma_1}{\alpha_2 - \alpha_1} \text{ and } n = \frac{\gamma_1 \cdot \alpha_2 - \gamma_2 \cdot \alpha_1}{\alpha_2 - \alpha_1}$$

Assuming an isotropic intensity distribution, the diffuse irradiance $E_{\text{diff,hor,A}i}$ through the area A_i on a horizontal surface becomes:

$$E_{\text{diff,hor,A}_i}(\mathbf{p_1},\mathbf{p_2}) = L_{\text{e,iso}} \int\limits_{\alpha_1}^{\alpha_2} \int\limits_{0}^{(m \cdot \alpha + n)} \sin\gamma \cos\gamma \, d\gamma \, d\alpha = \tfrac{1}{2} \cdot L_{\text{e,iso}} \int\limits_{\alpha_1}^{\alpha_2} \sin^2(m \cdot \alpha + n) \, d\alpha =$$

$$= \begin{cases} \tfrac{1}{2} \cdot L_{\text{e,iso}} \cdot (\alpha_2 - \alpha_1) \cdot \sin^2\gamma_1 & \text{for } m = 0 \\ \tfrac{1}{2} \cdot L_{\text{e,iso}} \cdot (\alpha_2 - \alpha_1) \cdot (\tfrac{1}{2} + \tfrac{1}{4} \cdot \dfrac{\sin 2\gamma_1 - \sin 2\gamma_2}{\gamma_2 - \gamma_1}) & \text{for } m \neq 0 \end{cases}$$

$$(2.38)$$

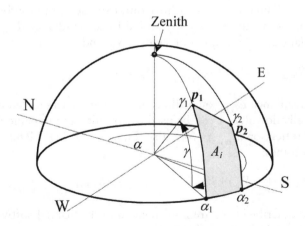

Figure 2.23 *Two Points, the Horizontal Meridian and Two Polar Meridians Define the Polygon Area*

The parameter $L_{e,iso}$ is the isotropic radiant intensity, that falls out of the equation when calculating the degree of diffuse shading. When using the anisotropic distribution, the integral also includes the angle-dependent radiant intensity L_e. However, only numerical methods are able to solve this integral. When simulating tilted planes, the angle of incidence θ, according to Equation (2.24), must be considered. Therefore, a complex analytical solution also exists (Quaschning and Hanitsch, 1998).

The diffuse irradiance $E_{diff,P}$ through the polygon surface of an object polygon with n polygon points $p_1 = (\alpha_1, \gamma_1)$ to $p_n = (\alpha_n, \gamma_n)$ can be calculated by:

$$E_{diff,P} = \left| \sum_{i=1}^{n-1} E_{diff,A_i}(\mathbf{p_i}, \mathbf{p_{i+1}}) \right| \tag{2.39}$$

Finally, the degree of *shading of diffuse irradiance* $S_{diff,hor}$ for a horizontal surface is the ratio of the diffuse irradiance reduction and the total diffuse irradiance. For an isotropic radiant intensity distribution, the shading becomes:

$$S_{diff,hor} = \frac{E_{diff,hor,P}}{E_{diff,hor}} = \frac{E_{diff,hor,P}}{\pi \cdot L_{e,iso}} \tag{2.40}$$

The degree of diffuse irradiance for a tilted surface is calculated analogously:

$$S_{diff,tilt} = \frac{E_{diff,tilt,P}}{E_{diff,tilt}} = \frac{E_{diff,tilt,P}}{L_{e,iso} \cdot \frac{\pi}{2} \cdot (1 + \cos \gamma_t)} \tag{2.41}$$

If the object polygon is transparent, the shading degree S must be weighted additionally with the transmittance τ.

Estimation of total shading

The global irradiance $E_{G,hor}$ on a horizontal surface can now be calculated using the direct irradiance $E_{dir,hor}$ and diffuse irradiance $E_{diff,hor}$ on the horizontal plane and the degrees of shading S_{dir} and $S_{diff,hor}$:

$$E_{G,hor} = E_{dir,hor} \cdot (1 - S_{dir}) + E_{diff,hor} \cdot (1 - S_{diff,hor}) \tag{2.42}$$

The global irradiance $E_{G,tilt}$ on a tilted surface can be calculated analogously. A corrected albedo value A can be used to consider any possible reduction of the ground reflection $E_{refl,tilt}$ (see also Equation 2.35). Hence, the global irradiance becomes:

$$E_{G,tilt} = E_{dir,tilt} \cdot (1 - S_{dir}) + E_{diff,tilt} \cdot (1 - S_{diff,tilt}) + E_{refl,tilt} \tag{2.43}$$

The methods described here are implemented in the SUNDI software that can be found on the CD-ROM of this book.

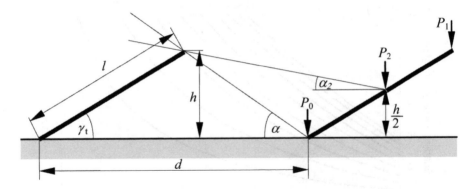

Figure 2.24 *Dimensions of Solar Energy Systems and Support Structure Rows*

Optimum distance of solar energy system support structures

Photovoltaic and solar thermal systems are often installed on the ground or on flat roofs. Support racks are used to mount the solar energy systems. Usually these are tilted to get higher annual irradiances, as explained in the section on irradiance on tilted surfaces (see p60).

Furthermore, horizontally installed solar energy systems have relatively high losses due to soiling, for example, through deposition of air pollutants, bird excrement or other dirt deposits on the surface of the collector. Rain or snow can clean tilted surface more easily. As a general rule, the lower the tilt angle the lower is the cleaning effect of rain and snow. In central European climates, average soiling losses in the range of 2–10 per cent can be expected for surfaces tilted at 30° if these are never cleaned manually. These losses can increase significantly for lower surface tilt angles. In other climatic regions with long periods without rain these losses can also increase significantly.

The disadvantage of rows of support structures is that they can cause self-shading. Optimization of the distance between the rows can decrease the shading losses. A way to estimate the optimum distance and tilt angle is described as follows.

The distance d between the rows and the length l of the solar energy system as shown in Figure 2.24 define the *degree of ground utilization*:

$$u = \frac{l}{d} \tag{2.44}$$

Shading will affect different positions on the tilted solar energy system surface differently. Shading has the highest occurrence at point P_0. If the utilization u increases, the shading losses will also rise.

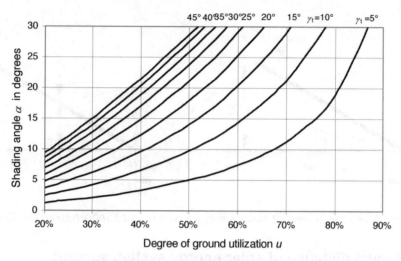

Figure 2.25 *Shading Angle α as a Function of the Degree of Ground Utilization* u *and the Surface Tilt Angle* γ_t

The *shading angle*

$$\alpha = \arctan\left(\frac{u \cdot \sin\gamma_t}{1 - u \cdot \cos\gamma_t}\right) \tag{2.45}$$

is a function of the ground utilization u and the surface tilt angle γ_t (see Figure 2.25). If the sun is directly in front of the solar energy system and the sun height is below the shading angle α, self-shading will occur.

With rising shading angles the irradiance losses increase. However, shading losses must be calculated individually for every location. The following calculations are made for Berlin but can easily be transferred to other locations with latitudes around 50° and with a central European climate. Locations at lower latitudes will have significantly reduced shading losses that often can be neglected.

Figure 2.26 shows the relative shading losses s at the point P_0 as a function of the shading angle α and the tilt angle γ_t. The annual irradiation $H_{G,tilt}$ on tilted but unshaded solar energy systems and the same annual irradiation $H_{G,tilt,red}$, reduced as a result of self-shading, define the *relative shading losses*:

$$s = 1 - \frac{H_{G,tilt,red}}{H_{G,tilt}} \tag{2.46}$$

It is apparent that solar energy systems are more sensitive to shading losses with higher surface tilt angles.

Since photovoltaic systems are very sensitive to shading, the irradiance at point P_0 could be used as a reference for the whole photovoltaic system. Table 2.13 shows the shading angle α and the resulting shading losses s for different degrees of surface utilization for tilt angles between 10° and 30°.

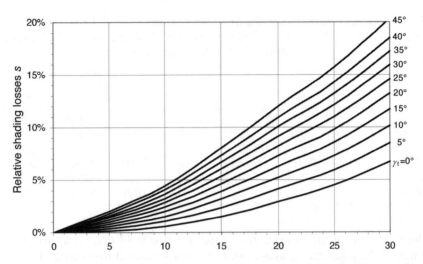

Figure 2.26 *Relative Shading Losses s as a Function of the Shading Angle α and Surface Tilt Angle γ$_t$ in Berlin (52.5°N)*

The gain factor *g* considers the irradiation gains due to the surface tilt angle (see also section on irradiance gain due to surface tilt or tracking, p64). The factor *g* is the ratio of the annual global irradiation $H_{G,tilt}$ on the tilted surface and the global irradiation $H_{G,hor}$ on a horizontal surface:

$$g = \frac{H_{G,tilt}}{H_{G,hor}} \tag{2.47}$$

The overall correction factor *c* considers the tilt gains and the shading losses. It is the ratio of the irradiation at point P_0 to that on a horizontal surface:

$$c = (1-s) \cdot g \tag{2.48}$$

Table 2.13 *Shading losses s, Gain Factor g and Overall Correction Factor c for Point P_0 at Different Ground Utilizations and Tilt Angles Calculated for Berlin (52.5°N)*

| | $\gamma_t = 30°$ | | | | $\gamma_t = 10°$ | | | |
u	α	s	g	c	α	s	g	c
1:1.5	38.8°	0.246	1.193	0.900	18.6°	0.048	1.088	1.036
1:2.0	23.8°	0.116	1.193	1.055	9.7°	0.015	1.088	1.072
1:2.5	17.0°	0.074	1.193	1.105	6.5°	0.009	1.088	1.078
1:3.0	13.2°	0.048	1.193	1.136	4.9°	0.006	1.088	1.081
1:3.5	10.7°	0.035	1.193	1.151	3.9°	0.004	1.088	1.084
1:4.0	9.1°	0.029	1.193	1.158	3.3°	0.004	1.088	1.084

Table 2.14 *Average Relative Shading Losses $\bar{s}$ and Overall Correction Factor c for Points* P_0, P_1 *and* P_2 *at Different Ground Utilizations and Tilt Angles Calculated for Berlin (52.5°N)*

	$\gamma_t = 30°$			$\gamma_t = 10°$		
u	$\bar{s}$	g	c	$\bar{s}$	g	c
1:1.5	0.098	1.193	1.076	0.018	1.088	1.068
1:2.0	0.048	1.193	1.136	0.006	1.088	1.081
1:2.5	0.032	1.193	1.155	0.004	1.088	1.084
1:3.0	0.021	1.193	1.168	0.003	1.088	1.085
1:3.5	0.016	1.193	1.174	0.002	1.088	1.086
1:4.0	0.013	1.193	1.177	0.002	1.088	1.086

The reduction of the ground utilization u below 0.33 (that is a ratio of the length of the solar energy system rows to the distance of the rows of 1:3) does not result in a significant reduction of the shading factor. Ground utilization higher than 0.4 (1:2.5) may be favoured by lower tilt angles. Considering the 5 per cent higher losses that result from soiling at a tilt angle of 10° instead of 30°, lower tilt angles are recommended for ground utilization values above 0.5 (1:2).

Solar thermal systems are less sensitive to shading than photovoltaic systems. Here, the average irradiance can be used for the estimation of the output of solar thermal systems. This is done by calculating the average shading losses at the points P_0, P_1 and P_2 shown in Figure 2.24. The losses at point P_0 have already been estimated. Point P_1 will never be shaded. For point P_2, the shading angle

$$\alpha_2 = \arctan\left(\frac{u \cdot \sin \gamma_t}{2 - u \cdot \cos \gamma_t}\right) \tag{2.49}$$

in the centre of the tilted surface is relevant. Table 2.14 shows the average relative shading losses $\bar{s}$ as well as the overall correction factor c for the three points P_0, P_1 and P_2. Compared to point P_0, the shading losses are much lower. No significant improvements are possible for ground utilization $u < 0.4$ (1:2.5). Lower tilt angles are generally not recommended.

Chapter 3

Solar Thermal Water Heating

INTRODUCTION

Solar thermal systems are an important use of solar energy. The use of solar thermal applications has a long history: Archimedes is reported to have boiled water using a concave mirror as early as 214 BC; thermal systems today still represent the most cost-effective use of solar energy.

Solar thermal signifies the thermal use of solar energy in general. Therefore, a number of different technical applications exist. Besides building space heating, water heating or industrial processes, solar thermal systems can be used for cooling applications or electricity generation with solar thermal power plants. The main operational areas are:

- solar swimming pool heating
- solar domestic water heating
- solar low-temperature heat for space heating in buildings
- solar process heat
- solar thermal electricity generation.

Since these operational areas are very far reaching, this chapter discusses only the main aspects of solar domestic water and swimming pool heating with closed and open solar collector systems. The following sections require the use of some thermodynamic quantities in the explanation of the principles. Table 3.1 summarizes the most important parameters and their symbols and units.

Table 3.1 *Thermodynamic Quantities for Thermal Calculations*

Name	Symbol	Unit
Heat, energy	Q	Ws (= J) or kWh
Heat flow	$\dot{Q}$	W
Temperature	ϑ	°C
Thermodynamic temperature	T	K (Kelvin, 0 K = –273.15°C)
Specific heat capacity	c	J/(kg K)
Thermal conductivity	λ	W/(m K)
Heat transition coefficient	k'	W/(m K)
Coefficient of heat transfer	k	W/(m² K)
Surface coefficient of heat transfer	α	W/(m² K)

Energy in form of *heat* Q is linked with the *heat flow* $\dot{Q}$

$$Q = \int \dot{Q} \, dt \tag{3.1}$$

Every temperature change $\Delta\vartheta$ also causes a *heat change* ΔQ. The change in heat can be calculated with the specific *heat capacity* c and the mass m of the affected material:

$$\Delta Q = c \cdot m \cdot \Delta\vartheta \tag{3.2}$$

Some confusion can occur owing to the use of different temperature scales. The Fahrenheit scale is generally not used in scientific work. However, the coexistence of the temperature ϑ given in degrees Celsius and the absolute temperature T given in Kelvin is problematic. The conversion of Celsius to Kelvin is given by:

$$T = \vartheta \cdot \tfrac{K}{°C} + 273.15 \text{ K} \tag{3.3}$$

The formula for the conversion of Fahrenheit to Celsius and Kelvin can be found in the appendix. The numerical value of the temperature difference $\Delta\vartheta$ in degrees Celsius (°C) is the same as the temperature difference ΔT in Kelvin (K). For a correct balance of units, the temperature difference in the equation above for the heat change should be given in Kelvin. The same goes for equations that are introduced in the following section. However, since the Celsius scale is more common than the Kelvin scale, the Celsius scale is used for most of the temperature differences and equations of this chapter.

The heat flow $\dot{Q}$ which causes the heat change for a constant heat capacity c, is:

$$\dot{Q} = \frac{dQ}{dt} = c \cdot \frac{dm}{dt} \cdot \Delta\vartheta + c \cdot m \cdot \frac{d\Delta\vartheta}{dt} \tag{3.4}$$

For heat capacities of various materials see Table 3.2.

Figure 3.1 shows a construction of n layers with a surface area A. On one side there is a temperature ϑ_1, on the other ϑ_2. This temperature gradient generates a heat flow through the layers given by:

$$\dot{Q} = k \cdot A \cdot (\vartheta_2 - \vartheta_1) \tag{3.5}$$

This heat flow $\dot{Q}$ causes a heat increase at the side with the lower temperature and a heat decrease on the other side until both sides have the same temperature. If the heat content of one side is much bigger than the other, the temperature change at the side with the high heat content can be neglected. For instance, the heat content of the environment surrounding a building is much higher than that inside the building. The heat flow through the building

Table 3.2 *Heat capacity* c *for Some Materials at* $\vartheta = 0$–$100°C$

Name	c in Wh/(kg K)	c in kJ/(kg K)	Name	c in Wh/(kg K)	c in kJ/(kg K)
Aluminium	0.244	0.879	Copper	0.109	0.394
Ice (–20°C to 0°C)	0.58	2.09	Air (dry, 20°C)	0.280	1.007
Iron	0.128	0.456	Brass	0.107	0.385
Ethanol (20°C)	0.665	2.395	Sand, dry	0.22	0.80
Gypsum	0.31	1.1	Water	1.163	4.187
Glass, glass wool	0.233	0.840	Heat transfer fluid		
Wood (spruce)	0.58	2.1	Tyfocor55% (50°C)	0.96	3.45

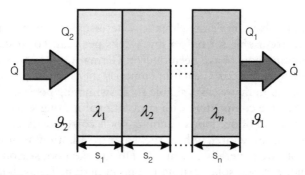

Figure 3.1 *Heat Transfer through* n *Layers with the Same Surface Area* A

walls does not change the outside ambient temperature, and this is true whether the ambient temperature is higher or lower than the building temperature. The *coefficient of heat transfer*

$$k = \left(\frac{1}{\alpha_1} + \frac{1}{\alpha_2} + \sum_{i=1}^{n} \frac{s_i}{\lambda_i} \right)^{-1} \tag{3.6}$$

can be calculated with the surface coefficient of heat transfer α_1 and α_2 of both sides, the thermal conductivity λ_i and the layer thickness s_i of all n layers. Table 3.3 shows the heat conductivity λ of various materials.

SOLAR THERMAL SYSTEMS FOR WATER HEATING

Solar thermal swimming pool heating

This chapter first deals with swimming pool heating. This is not because heated swimming pools have any ecological advantages – they always draw a high demand on drinking-quality water and energy. However, the low-temperature heat demand for pool heating allows the use of simple and economic solar energy systems, which have seen widespread deployment in this sector.

Table 3.3 *Thermal Conductivity of Various Materials*

Material	Heat conductivity λ in W/(m K)	Material	Heat conductivity λ in W/(m K)
Steel	60	Cork	0.045
Iron	81	Sheep's wool	0.040
Ice (0°C)	2.23	Glass wool	0.040
Reinforced concrete	1.80	Rock wool	0.040
Glass	0.81	Polystyrene (PS)	0.034
Water (20°C)	0.60	Polyurethane (PUR)	0.030
Gypsum	0.45	Air (dry, 20°C)	0.026
Wood (spruce)	0.14	Glass fibre vacuum	0.003

Swimming pools in moderate climatic zones usually need heating systems; otherwise they are usable for only a few weeks per year. For instance, about 500,000 swimming pools have been built in Germany. Since average ambient temperatures are below 20°C even in summer, there is a huge potential for solar pool heating. In many cases simple solar swimming pool heating systems have already become competitive with conventional heating systems.

The heat demand of outdoor swimming pools corresponds very well to the solar irradiation. In winter, when the solar irradiation is low, outdoor swimming pools are generally not in use. During the pool season in summer and transitional periods, solar heating is a good option. Today, huge amounts of fossil fuels are wasted in outdoor pool heating, although solar pool heating as shown in Figure 3.2 could replace most fossil fuel-based heating systems.

Water temperatures of exposed areas in central European climates are usually between 16°C and 19°C during the pool season. A temperature rise of a few degrees is normally sufficient for comfort. For such a low-temperature heat demand, simple absorbers can be used. These absorbers convert solar radiation to available heat for the swimming pool. A pump moves the pool water through the absorber. The absorber heats up the water and returns it to the pool. A hot water storage tank such as is common in domestic water heating systems is not needed – the pool water itself serves as heat storage.

If a dark-coloured hose is laid onto the solar irradiated ground in summer, the water inside the hose becomes hot in a relatively short time. A swimming pool absorber is not much more complicated. It can be made of black tubes that are installed permanently on large surfaces such as roofs.

The *absorber material* is usually plastic. However, the material must be resistant to degradation caused by ultraviolet sunlight and chlorinated pool water. Some suitable materials are polyethylene (PE), polypropylene (PP) and ethylene propylene diene monomer (EPDM). EPDM has a longer lifetime but also costs more. PVC should not be used for ecological reasons – it can emit highly toxic dioxins if it burns.

The *pump* of the system should only operate if the absorber can achieve a temperature rise of the pool water. If the pump operates under very cloudy

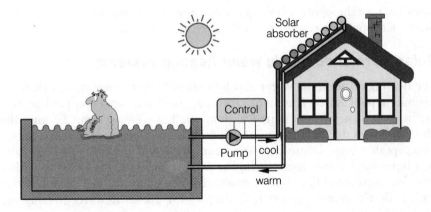

Figure 3.2 *Principle of Solar Thermal Swimming Pool Heating*

conditions or during the night, the pool water cools down in the absorber, which now acts as a radiator. A simple two-step controller with hysteresis can avoid this; if sensors detect that the temperature difference between pool and absorber is above a certain threshold level, the pump is switched on.

A conventional auxiliary heating system can be integrated to ensure that the desired pool temperature can always be achieved. If the pool is solar heated only, the water temperature will fluctuate with the weather. During bad weather periods, the water temperature is slightly lower; however, this is often acceptable because the pool is not used so much under these conditions.

The typical *heat demand of outdoor swimming pools* in moderate climates is between 150 kWh and 450 kWh per square metre of pool surface. A well-designed solar heating system can maintain a base temperature of 23°C, and thus a fossil heating system is not necessary. For a pool with a surface area of 2000 m², a solar heating system can avoid the burning of 75,000 litres of fuel oil and the production of 150,000 kg CO_2 (boiler efficiency 80 per cent) every season. Covering the pool during the night can minimize the heat losses and save additional energy.

As a rule of thumb, the *size of the solar absorber surface* should be 50–80 per cent of the pool surface; however, this depends significantly on the climate. Experience from previous installations or computer simulations can provide more exact values for the system designer. The absorber costs are of the order €100/m². Usually the costs of solar heating systems are lower than the costs of fossil fuel heating systems. Only if the outdoor swimming pool is to be operated all year round or if the pool temperature has to be rather high would an auxiliary fossil heating system reduce costs compared to solar energy solutions alone.

Electricity is needed to run the pump of a solar swimming pool heating system (see also Chapter 6, section on costs of a solar thermal system for domestic water heating, p240); however, a small photovoltaic system could generate this electricity. Then, the temperature control can be omitted in some cases because the photovoltaic generator runs the pump only when the sun shines. Chapter 4 describes photovoltaic systems in detail.

Solar thermal domestic water heating systems

The heating of domestic water involves much higher temperatures than for swimming pool water. The simple absorbers used for swimming pool heating are in most locations not suited to domestic hot water systems because the absorber losses due to convection, rain and snow as well as heat radiation are unacceptably high. Domestic water heating systems typically use collectors that have much lower losses at higher water temperatures. These are either flat-plate, evacuated flat-plate or evacuated tube collectors and are integrated with collector storage systems. Collectors for domestic water heating are described in the section on solar collectors, p85.

A complete system for domestic water heating consists not only of a closed collector to heat the water. Further components such as a hot water storage tank, pump and an intelligent control unit are needed to ensure a hot water supply that is as comfortable as we expect from conventional systems.

A very *simple system for solar water heating* can be made of a black water-filled tank that is exposed to sunlight in summer. If the tank is installed higher than the tap, the warm water can be used without any further component. An example for such an application is a solar shower that is sold as camping equipment. In principle, it is a black sack hung on a high branch of a tree. If this sack is exposed for some hours to solar radiation, a shower with solar heated water can be taken.

However, this system does not meet the demand of daily routine. After the sack is empty it must be refilled again by hand. To avoid this inconvenience, sack and tap can be pressure-sealed and a hose can then be connected to replace water automatically. As a further improvement a solar collector with a high efficiency all year round can replace the sack. However, the collector content is only sufficient for a very short shower and the water temperatures will be very high. Therefore, a storage tank is needed. Two systems to integrate hot water storage tanks into solar energy systems are described in the following sections.

Thermosyphon systems

A thermosyphon system as shown in Figure 3.3 makes use of gravity. Cold water has a higher specific density than warm water. It is therefore heavier and sinks to the bottom. The collector is always mounted below the water storage tank. Cold water from the bottom of the storage tank flows to the solar collector through a descending water pipe. When the collector heats up the water, the water rises again and flows back to the tank through an ascending water pipe at the upper end of the collector. The cycle of tank, water pipes and collector heats up the water until temperature equilibrium is reached. The consumer can draw off hot water from the top of the tank. Used water is replaced through a fresh supply of cold water through an inlet at the bottom

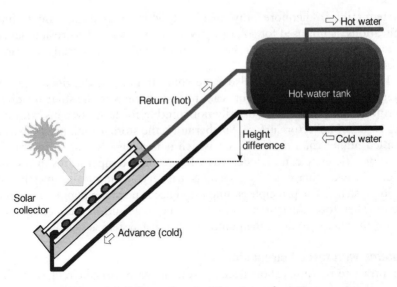

Figure 3.3 *Schematic of a Thermosyphon System*

of the tank. This cold water joins the cycle and is heated in the collector in the same way as before. Due to higher water temperature differences at higher solar irradiances, the warm water rises faster than at lower irradiances and the flow rates are increased. Therefore, the water circulation adapts itself nearly perfectly to the available solar irradiance.

It is very important that the storage tank of a thermosyphon system is well above the collector; otherwise the cycle can run backwards at night and cool the water from the storage tank over the collector. In regions with high solar irradiation and flat-roof architecture, storage tanks are usually put on the roof. The collector is also mounted on the roof or on the wall of the sunny side of the house.

With gable roofs, the storage tank must be mounted as high as possible under the roof if the collector is also installed on the roof. The high weight of the water-filled tank can sometimes cause structural problems. Furthermore, integration with a conventional heating system, which is usually placed in the basement, is more difficult.

A system where the water flows directly through the collector is called a *single-circuit system*. Such a system is only suitable for frost-free regions; otherwise the water can freeze in the collector and pipes and destroy the system. In regions with the possibility of frost, a *double-circuit system* is frequently used, in which the water is kept inside the storage tank. A second quantity of water is mixed with an antifreeze agent to use as a working fluid in the solar cycle. A heat exchanger transfers the heat from the solar cycle to the storage tank, thus separating the usable water from the antifreeze mixture. Glycols are often used as antifreeze agents; however, antifreeze agents should be non-toxic because they can contaminate the hot water supply in the case of a system failure. Therefore, ethylene glycol, which is used for many technical applications, is not used for solar energy systems. To avoid corrosion damage, the antifreeze agent must also be compatible with the materials used in the system construction.

Thermosyphon systems also have some important disadvantages. The system itself is inert and cannot react to fast changes in the solar irradiance. Thermosyphon systems are usually not suitable for large systems with more than 10 m² of collector surface. Furthermore, the storage tank must always be installed higher than the collector, which is not always easy to realize. The collector efficiency can also decrease due to high temperatures in the solar cycle. However, thermosyphon systems are very economical domestic water heating systems. The principle is simple and needs neither a pump nor a control system. Therefore, the system cannot fail due to a fault in these components. Finally, the energy to drive the pump and control system is saved.

Systems with forced circulation

In contrast to thermosyphon systems, systems with forced circulation use an electrical pump to move the water in the solar cycle. The collector and storage tank can be installed independently and a height difference between the tank and collector is no longer necessary. However, the pipe lengths should be designed to be as short as possible since all warm water pipes cause heat losses. Figure 3.4 shows a system with forced circulation.

Two *temperature sensors* monitor the temperatures in the solar collector and the storage tank. If the collector temperature is above the tank temperature by a certain threshold, the control starts the pump. The pump moves the heat transfer fluid in the solar cycle. The switch-on temperature difference is normally between 5 and 10°C. If the temperature difference decreases below a second threshold, the control switches the pump off again. The choice of both thresholds must ensure that the pump does not continually switch on and off during low irradiance conditions.

Conventional circulation *pumps* made for heating installations can be used for the solar cycle. These pumps are reliable and economic. Most pumps have various velocity stages to adapt the flow rate to the solar irradiance. Pumps are usually designed for flow rates of 30–50 litre/h per square metre of solar collector area. Higher flow rates are chosen for swimming pool absorbers since the temperature requirement is lower and the water needs less heating. However, if the flow rate is too low, the temperature in the collector rises and the system efficiency decreases. On the other hand, if the flow rate is too high, the energy demand to drive the pump is unnecessarily high.

The pump usually runs at the alternating voltage of the public grid. It is also possible to use DC motors to drive the pump. A small photovoltaic system can provide the electrical energy needed. In that case, all of the energy for the system comes from the sun.

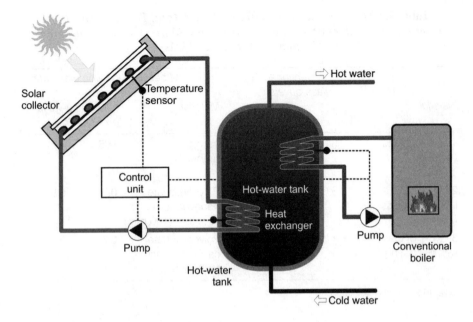

Figure 3.4 *Schematic of a Double-Cycle System with Forced Circulation*

SOLAR COLLECTORS

To obtain the higher temperatures required for domestic water heating, the following collector types are available:

- integral storage collector systems
- flat-plate collectors
- evacuated flat-plate collectors
- evacuated tube collectors.

The flat-plate collectors and evacuated tube collectors always need a storage tank. The collector water volume itself is very low and water temperatures inside the collector can reach much more than 100°C if the water is not continuously replaced. The collector efficiency also decreases significantly at high operating temperatures. Therefore, all solar collector systems for domestic water heating need a hot water storage tank to store larger quantities of absorbed collector heat and to provide a hot water supply during bad weather periods and darkness. In contrast, for solar swimming pool systems, the pool water itself is the storage.

Integral collector storage systems

For integral collector storage systems (ICS) the hot water storage tank is integrated with the collector itself. A technically robust system is not easy to

Table 3.4 *Heat Transition Coefficient* k *and Total Energy Transition Coefficient (g-value) of Various Conventional Materials and Transparent Insulation Materials (TIMs)*

Material	Conventional glass covers		Material	TIM between low-iron glass with air gap	
	k in W/(m² K)	g-value		k in W/(m² K)	g-value
1 layer of glass (4 mm)	5.9	0.86	Aero gel granulate (20 mm)	0.85	0.4
2 layers of insulating glass (20 mm)	3.0	0.77	Polycarbonate honeycomb structure (100 mm)	0.7	0.66
3 layers of glass with IR coating (36 mm)	1.0–1.2	0.53–0.62	Polycarbonate capillary structure (100 mm)	0.7	0.64

Note: IR = infrared reflecting

construct. If the system is used in regions with a danger of frost, heat losses cool both the collector and storage; ultimately, they can freeze and be damaged. A double-circuit system with frost protection is not possible to realize as an integral collector storage system. A way needs to be found for achieving significant reduction of the collector heat losses. Better insulation on the back is not a problem – the problem is heat losses through the front cover. Sunlight must pass through the front with low absorption and reflection losses. The cover must therefore be transparent, and yet this leads to large heat losses through the cover. A vacuum can reduce the heat losses, but not as much as is necessary to design an integral collector storage system.

New so-called *transparent insulation materials* (TIM) brought a solution to these problems (Lien et al, 1997; Manz et al, 1997). These materials have a slightly lower transmittance compared to low-iron solar safety glass. However, the heat transition coefficient is significantly lower so that the heat losses are reduced to levels acceptable for ICS systems. Table 3.4 compares various conventional and TIM covers.

Figure 3.5 shows a sketch of an ICS system. The hot water tank is made of stainless steel. The back is perfectly insulated. Reflectors on the inner side of the back reflect the light to the storage, which is also the absorber of this system. The transparent insulation material is under the glass front cover. A system covering two square metres has a storage volume of about 160 litres.

ICS systems do not need an external heat storage tank, which is necessary for other collector systems. The total system is simpler: some components found in other collector systems are not necessary and this reduces the cost. If the water temperature in the storage tank is too low, an auxiliary, thermostat-controlled water heater can boost the temperature up to the desired level.

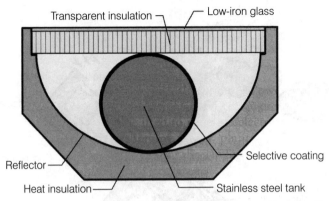

Figure 3.5 *Cross-section through an Integral Collector Storage System*

Disadvantages of the ICS systems are the high weight and large dimensions. This makes installation more difficult in many cases. Furthermore, the system efficiency is usually lower than that of an optimal system with forced circulation. These are some of the reasons why the integral system type has not reached a high market penetration to date.

Flat-plate collectors

The most common collectors for solar domestic water heating systems in many countries today are flat-plate collectors. These mostly consist of three components:

- transparent cover
- collector housing
- absorber.

An absorber is inside the flat-plate collector housing. This absorber converts sunlight to heat and transfers it to water in tubes, which passes through the system. The collector housing is highly insulated on the back and sides to keep heat losses to a minimum; however, there are still some collector heat losses, which mainly depend on the temperature difference between the absorber and ambient air. These losses are subdivided into convection and radiation losses. Movements of the air cause the convection losses.

A pane of glass covers the collector and avoids most of the convection losses. Furthermore, it reduces heat radiation from the absorber to the environment in the same way as a greenhouse. However, the glass also reflects a small part of the sunlight that can no longer reach the absorber. Figures 3.6 and 3.7 show the mechanism and energy flow in flat-plate collectors.

The *front glass cover* reflects and absorbs a small part of the solar radiant power Φ_e as shown in Figure 3.8; however, the majority of the solar radiation passes through the glass. The reflectance ρ, absorptance α and transmittance τ

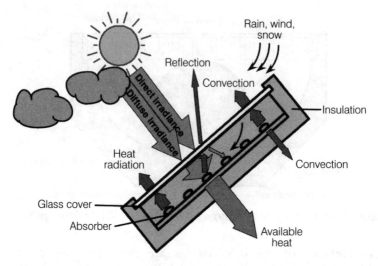

Figure 3.6 *Processes in a Flat-plate Collector*

can describe these processes. The sum of these three values must always be equal to one:

$$\rho + \alpha + \tau = 1 \qquad (3.7)$$

The corresponding radiant powers are:

$$\Phi_e = \Phi_\rho + \Phi_a + \Phi_\tau = \rho \cdot \Phi_e + \alpha \cdot \Phi_e + \tau \cdot \Phi_e \qquad (3.8)$$

The absorption of the solar radiation heats up the pane of glass. If the glass is in thermal equilibrium, it must emit the absorbed radiation. Then, the emitted radiant power Φ_e is equal to the absorbed radiant power Φ_a, otherwise the glass would heat up indefinitely. Hence, the emittance ε is equal to the absorptance α:

$$\alpha = \varepsilon \qquad (3.9)$$

On one hand, the front cover should let through the majority of the solar radiation. On the other hand, it should also keep back the heat radiation of the absorber and reduce the convection losses to the environment. Most collectors use a single glass layer made of thermally treated low-iron solar glass. This glass has a high transmittance ($\tau \rightarrow 1$) and a good resistance to the influences of the environment. Front covers made of glass prevail against those made of plastic because the lifetime of plastic is limited due to a poorer resistance to ultraviolet radiation and the influences of weather.

Double glazing can result in further heat loss reductions but also reduces the transmitted solar radiant power and increases costs.

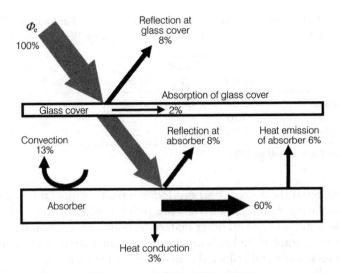

Source: Wagner and Co, 1995

Figure 3.7 *Energy Conversion in the Solar Collector and Possible Losses*

The use of special materials for the front cover can increase the collector efficiency. These materials should let incoming radiation through, but reflect the outgoing infrared emissions coming from the absorber plate back inside. Infrared reflecting glass such as In_2O_3 or ZnO_2 with high transmittance for visible light but high reflectance for infrared meets these requirements. Table 3.5 shows the parameters of these materials; however, higher costs and poorer transmittance of visible light compared to standard glass have prevented the widespread use of these materials.

The *collector housing* can be made of plastic, metal or wood. The housing must seal the front glass cover so that no heat can escape and no dirt, animals or humidity can get into the collector. Many collectors have controlled ventilation to avoid humidity inside the collectors condensing on the inside of the cover glass.

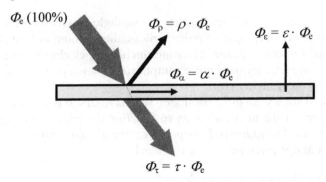

Figure 3.8 *Processes at the Collector Front Glass Cover*

Table 3.5 *Absorption, Transmission and Reflection Factors for IR Glass*
In_2O_3 and ZnO_2 Compared with Ordinary Window Glass

Material	Visible			Infrared		
	$\alpha = \varepsilon$	τ	ρ	$\alpha = \varepsilon$	τ	ρ
Window glass	0.02	0.97	0.01	0.94	0	0.06
In_2O_3	0.10	0.85	0.05	0.15	0	0.85
ZnO_2	0.20	0.79	0.01	0.16	0	0.84

Source: Kleemann and Meliß, 1993

The material for the *rear heat insulation* must be temperature resistant and highly insulating. Suitable materials are polyurethane foam or mineral fibre slabs. The insulation and all other materials should not contain binding agents that may evaporate at higher temperatures because these could condense on the front glass pane and reduce the incoming solar radiation.

The collector can reach stagnation temperatures of about 200°C. All materials used must therefore resist these temperatures. As a result, the absorber is usually made of metal such as copper, steel or aluminium. Various absorber designs are used as shown in Figure 3.9. Absorbers with soldered or forced copper pipe are the most commonly used today. Aluminium is now little used as absorber material, since it is not corrosion-resistant and its production needs much more energy input than other materials.

It is well known that black materials absorb sunlight very well and warm up to higher temperatures. However, metallic materials do not naturally have black surfaces and must therefore be coated. Black lacquer is one option. Temperature-resistant lacquer serves its purpose but there are more advanced materials available for *absorber coating*. If a black surface heats up, it re-emits part of the absorbed heat energy as heat radiation. This can be noticed with electrical hotplates for cooking. The heat radiation can be felt on the skin without touching the hotplate. A black lacquered absorber exhibits the same effect. It transfers only a part of the absorbed heat to the water flowing through; the rest of the heat is undesirably emitted as heat radiation to the environment.

So-called *selective coatings* absorb the sunlight nearly as well as black lacquered surfaces but re-emit a significantly smaller amount of heat radiation. Materials used for these advanced coatings include black chrome, black nickel or TiNOX. However, they need more complicated coating processes than black lacquer and so incur higher costs.

Figure 3.10 shows the different material behaviours. It is necessary to look at the spectrum of the heat radiation to describe the principle of selectively coated absorbers. The *radiance L_e* depends on the absolute temperature T and wavelength λ and is given according to Planck by:

$$L_{e,\lambda}(\lambda,T) = \frac{c_1}{\lambda^5} \cdot \frac{1}{\exp\left(\dfrac{c_2}{\lambda \cdot T}\right) - 1} \cdot \frac{1}{\Omega_0}$$

(3.10)

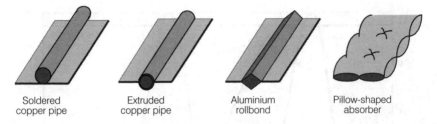

| Soldered copper pipe | Extruded copper pipe | Aluminium rollbond | Pillow-shaped absorber |

Figure 3.9 *Various Designs of Solar Absorber*

The constants c_1 and c_2 are:

$$c_1 = 2 \cdot h \cdot c^2 = 1.191 \cdot 10^{-16} \text{ Wm}^2 \tag{3.11}$$

$$\text{and } c_2 = \frac{h \cdot c}{k} = 1.439 \cdot 10^{-2} \text{ mK} \tag{3.12}$$

$\Omega_0 = 1$ sr is needed for the correct balance of the units. The steradian (sr) is the unit of the solid angle. The sun can ideally be considered as a black body with a temperature of 5777 K.

The majority of the solar spectrum is in the wavelength range below 2 μm. The absorber should have a very high absorptance in this range. The sun heats up the absorber to about 350 K. The maximum of the corresponding emittance spectrum is higher than 2 μm. Since the absorptance α is identical to the emittance ε according to Kirchhoff's law of emission of radiation, the absorptance should be as low as possible in the range above 2 μm so that the heated absorber emits only a little heat radiation to the environment.

Figure 3.11 shows the relative radiance referred to the maximum of the corresponding spectra at 350 K and 5777 K as well as the absorptance of a selective and non-selective absorber. Table 3.6 summarizes parameters of selective and non-selective absorber materials. Many collectors that are sold today use selective coatings such as TiNOX (TiNOX, 2004).

A *vacuum* between the front glass cover and the absorber can significantly reduce convection heat losses due to air movements inside the collector. The

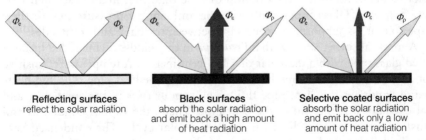

| **Reflecting surfaces** reflect the solar radiation | **Black surfaces** absorb the solar radiation and emit back a high amount of heat radiation | **Selective coated surfaces** absorb the solar radiation and emit back only a low amount of heat radiation |

Figure 3.10 *Losses at Absorber Surfaces with Different Types of Coating*

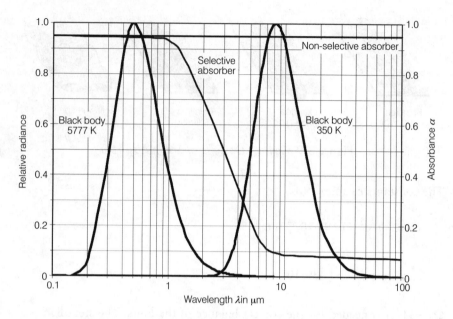

Figure 3.11 *Spectra of Black Bodies at 5777 K and 350 K and the Absorptance of Selective and Non-selective Absorbers*

evacuated flat-plate collector uses this principle. Since the ambient air pressure might press the front cover against the absorber, small supports between the back of the collector and the front cover must keep the front cover in position. However, it is difficult to maintain the vacuum over a long period of time. Ambient air nearly always finds a path between the glass and collector housing to get into the collector airspace. Therefore, an evacuated flat-collector must be evacuated from time to time with a vacuum pump, which is connected to a valve at the collector. The evacuated tube collector, which is described in the next section, can avoid these disadvantages.

Evacuated tube collectors

The high vacuum inside the closed *glass tube* of the evacuated tube collector is easier to maintain over a long period of time than that in an evacuated flat-plate collector. Glass tubes can resist the ambient air pressure due to their shape so that no supports are necessary between the back and front sides.

A metal absorber sheet with a *heat pipe* in the middle is embedded inside a closed glass tube with a diameter of a few centimetres. A temperature sensitive working medium such as methanol is used inside the heat pipe. The sun heats up and vaporizes this heat pipe fluid. The vapour rises to the condenser and heat exchanger at the end of the heat pipe. There, the vapour condenses and transfers the heat to the heat carrier of the solar cycle. The condensed heat pipe fluid flows back to the bottom of the heat pipe where the sun starts heating it again. To work properly, the tubes must have a minimum angle of

Table 3.6 *Absorptance* α, *Transmittance* τ *and Reflectance* ρ *for Different Absorber Materials*

	Visible			Infrared		
Material	α = ε	τ	ρ	α = ε	τ	ρ
Non-selective absorber	0.97	0	0.03	0.97	0	0.03
Black chrome	0.87	0	0.13	0.09	0	0.91
Black nickel	0.88	0	0.12	0.07	0	0.93
TiNOX (TiN + TiO + TiO$_2$)	0.95	0	0.05	0.05	0	0.95

inclination to allow the vapour to rise and the fluid to flow back. A cross-section illustrating the operating principle of the evacuated tube collector is shown in Figure 3.12. The photo in Figure 3.13 shows a detail of an evacuated tube collector.

There are also evacuated tube collectors with a heat pipe that passes through the end of the glass tube. The heat transfer medium of the solar cycle can flow directly through the heat pipe of these collectors. Then, a heat exchanger is no longer needed and the collector need not be mounted with a minimum angle of inclination.

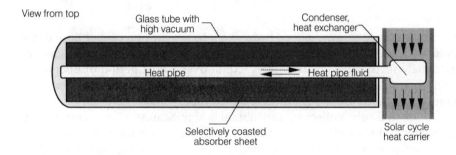

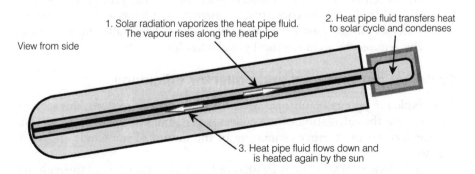

Figure 3.12 *Assembly and Function of the Evacuated Tube Collector with Heat Pipe*

Figure 3.13 *Photo of the Connections of the Evacuated Tubes to the Solar Cycle*

The entrance of atmospheric hydrogen into the vacuum cannot be avoided even with nearly perfectly closed glass tubes because hydrogen atoms are extremely small. This destroys the vacuum over a long period of operation. Therefore, so-called getters, which can absorb hydrogen over a long period of time, are installed inside the evacuated glass tube.

Evacuated tube collectors can obtain a significantly higher energy gain especially in the cooler months of the year. Thus, a solar thermal system with evacuated tube collectors needs a smaller collector area compared to standard flat-plate collectors. The specific collector prices of evacuated tube collectors are higher than that of flat-plate collectors. Furthermore it is not possible to integrate tube collectors directly into a roof; they must always be installed on top of the roof. This reduces the architectonical possibilities for collector integration.

Collector performance and collector efficiency

After explaining the assembly of collectors in the previous sections, this section will describe the calculation of the collector performance and efficiency. This is necessary for the further estimation of the output of the whole collector system.

The collector converts solar irradiance E, which is transmitted through the front glass cover with transmittance τ onto the collector surface A_C, directly to heat. The power output of the solar collector $\dot{Q}_{out}$ is reduced by losses due to reflection $\dot{Q}_{ref}$, convection $\dot{Q}_{conv}$ and heat radiation $\dot{Q}_{rad}$

$$\dot{Q}_{out} = \tau \cdot E \cdot A_C - \dot{Q}_{ref} - \dot{Q}_{conv} - \dot{Q}_{rad} \qquad (3.13)$$

The convection losses $\dot{Q}_{conv}$ and heat radiation losses $\dot{Q}_{rad}$ can be combined as $\dot{Q}_{RC}$. The heat radiation losses $\dot{Q}_{rad}$ of selective absorbers are much lower than the radiation losses of non-selective absorbers as described above. A vacuum between the front cover and the absorber can reduce the convection losses $\dot{Q}_{conv}$ as described for the evacuated flat-plate and tube collector. The reflection losses $\dot{Q}_{ref}$ can be estimated using the reflectance ρ from the irradiance passing the glass front cover.

With

$$\dot{Q}_{RC} = \dot{Q}_{conv} + \dot{Q}_{rad}$$

and

$$\dot{Q}_{ref} = \tau \cdot \rho \cdot E \cdot A_C$$

and the collector power output becomes:

$$\dot{Q}_{out} = \tau \cdot E \cdot A_C \cdot (1 - \rho) - \dot{Q}_{RC} \qquad (3.14)$$

Using the absorptance $\alpha = 1 - \rho$ of the absorber, the equation reduces to:

$$\dot{Q}_{out} = \tau \cdot \alpha \cdot E \cdot A_C - \dot{Q}_{RC} = \eta_0 \cdot E \cdot A_C - \dot{Q}_{RC} \qquad (3.15)$$

with $\quad \eta_0 = \alpha \cdot \tau \qquad (3.16)$

η_0 is called the *optical efficiency*. It describes the collector efficiency without any losses due to convection or heat radiation. This is only the case if the absorber temperature is equal to the ambient temperature.

The thermal losses Q_{RC} depend on the collector temperature ϑ_C and the ambient temperature ϑ_A as well as on the coefficients a, respectively, a_1 and a_2:

$$\dot{Q}_{RC} = a_1 \cdot A_C \cdot (\vartheta_C - \vartheta_A) + a_2 \cdot A_C \cdot (\vartheta_C - \vartheta_A)^2 \approx a \cdot A_C \cdot (\vartheta_C - \vartheta_A) \qquad (3.17)$$

Table 3.7 shows the optical efficiencies η_0 as well as the loss coefficients a_1 and a_2 for various collectors. The loss coefficients of evacuated tube collectors are much lower than those of non-evacuated flat-plate collectors, hence their higher efficiency at low ambient temperatures or low irradiances. The loss coefficients are usually estimated from collector tests. Some calculations only use the single loss coefficient a instead of the two loss coefficients a_1 and a_2. However, in that case it is not possible to eliminate only a_2 since that can cause

Table 3.7 *Optical Efficiencies η_0 and Loss Coefficients a_1 and a_2 of Real Collectors with the Collector Absorber Area A_C as Reference*

Name	Type	η_0	a_1 in W/(m² K)	a_2 in W/(m² K²)	A_C in m²
Paradigma Solar 500	Flat-plate	0.805	3.79	0.009	4.7
Solahart M	Flat-plate	0.746	4.16	0.0084	1.815
Solahart OYSTER Ko	Flat-plate	0.803	2.49	0.0230	1.703
Sonnenkraft SK 500	Flat-plate	0.800	3.02	0.0013	2.215
Wagner Euro C18 Microtherm Sydney	Flat-plate	0.789	3.69	0.007	2.305
SK-6	Evacuated tube	0.735	0.65	0.0021	0.984
Thermolux 2000-6R	Evacuated tube	0.801	1.13	0.008	1.05
Ritter CPC 12 OEM	Evacuated tube	0.617	1.04	0.0013	2.01
Sunda SEIDO 5-16	Evacuated tube	0.736	1.78	0.0130	2.592

Source: SPF Institut für Solartechnik, 2002

high errors. The coefficient *a* must be estimated separately from measurements. The collector reference surface A_C must always be given with the collector parameters. It is possible to determine the collector efficiency parameters referring to the absorber area, aperture area or total collector area. The absorber area is used as a reference for the following calculations.

The *collector efficiency* η_C can be calculated using the power output of the solar collector $\dot{Q}_{out}$ as well as the solar irradiance E, which reaches the collector surfaces A_C.

With $\eta_C = \dfrac{\dot{Q}_{out}}{E \cdot A_C} = \eta_0 - \dfrac{\dot{Q}_{RC}}{E \cdot A_C}$, the *collector efficiency* becomes:

$$\eta_C = \eta_0 - \frac{a_1 \cdot (\vartheta_C - \vartheta_A) + a_2 \cdot (\vartheta_C - \vartheta_A)^2}{E} \approx \eta_0 - \frac{a \cdot (\vartheta_C - \vartheta_A)}{E} \qquad (3.18)$$

Figure 3.14 shows the typical collector efficiencies of a flat-plate collector. The thermal losses increase with rising temperature differences between the collector and ambient air. At low solar irradiances the efficiency decreases faster. For instance, at a solar irradiance of only 200 W/m² the output of the collector of this example becomes zero at a temperature difference of about 40°C.

The *stagnation temperature* of the collector can also be found from the figure. The stagnation temperature is the temperature at which the collector power output and the collector efficiency are equal to zero ($\eta_C = 0$). At an irradiance of 400 W/m², the stagnation temperature of the collector is about 75°C above the ambient temperature. The stagnation temperature can rise above 200°C at irradiances of 1000 W/m². Therefore, collector materials must be chosen that can resist these relatively high temperatures over a long period of time.

The calculation of the collector efficiency given above is only valid if there is no wind; convective thermal losses will increase with the wind speed.

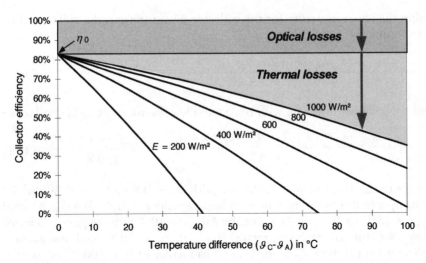

Figure 3.14 *Collector Efficiencies* η_C *at Different Irradiances* E *and Temperature Differences* $\Delta\vartheta$

Modifications of the loss factors can consider this. The optical efficiency η_0 also depends on the angle of incidence and the solar spectrum. For very exact calculations these effects must be also taken into account.

Pipes

A system for solar water heating also requires pipes. These pipes transport the heat transfer medium from the solar collector to the heat storage unit and the water from the storage unit to the consumer.

From an energy point of view, pipes incur undesirable losses. They do not play a role in room heating except for a short period in winter during which they may provide a very small contribution to the overall demand. If the pipes are badly insulated or not insulated at all, they waste the majority of the collected heat on its way to the consumer.

The collector flow rate, i.e. the mass of heat transfer medium that flows through the collector per hour, is influenced by the lengths and diameters of the pipes. On one hand, the collector flow rate should be as high as possible so that the collector efficiency does not decrease due to high collector temperatures. On the other hand, the flow rate should not be too high so as to keep the pumping energy demands within an acceptable range.

The *collector flow rate* $\dot{m}$ can be calculated using the collector power output $\dot{Q}_{out}$, the heat capacity c of the heat transfer medium and the desired temperature difference $\Delta\vartheta_{HTF}$ of the heat transfer fluid between collector inlet and outlet:

$$\dot{m} = \frac{\dot{Q}_{out}}{c \cdot \Delta\vartheta_{HTF}} \qquad (3.19)$$

Using the above equation for the collector power output, the collector flow rate becomes:

$$\dot{m} = \frac{\eta_0 \cdot E \cdot A_C - a_1 \cdot A_C \cdot (\vartheta_C - \vartheta_A) - a_2 \cdot A_C \cdot (\vartheta_C - \vartheta_A)^2}{c \cdot \Delta \vartheta_{HTF}} \tag{3.20}$$

and the collector flow rate with respect to the collector or absorber surface is:

$$\dot{m}' = \frac{\dot{m}}{A_C} = \frac{\eta_0 \cdot E - a_1 \cdot (\vartheta_C - \vartheta_A) - a_2 \cdot (\vartheta_C - \vartheta_A)^2}{c \cdot \Delta \vartheta_{HTF}} \approx \frac{\eta_0 \cdot E - a \cdot (\vartheta_C - \vartheta_A)}{c \cdot \Delta \vartheta_{HTF}} \tag{3.21}$$

For instance, if a flat-plate collector with $\eta_0 = 0.8$ and $a = 4$ W/(m^2 K) heats up a heat transfer medium with heat capacity $c = 0.96$ Wh/(kg K) from $\vartheta_{Cin} = 35°C$ to $\vartheta_{Cout} = 45°C$ (i.e. by $\Delta \vartheta_{HTF} = 10$ K), the required collector flow rate for an ambient temperature of $\vartheta_A = 20°C$ and an average collector temperature $\vartheta_C = 40°C$ at an irradiance of $E = 800$ W/m^2 is $\dot{m}' = 58.3$ kg/(m^{2}h).

Replacing

$$\Delta \vartheta_{HTF} = \vartheta_{Cout} - \vartheta_{Cin} \tag{3.22}$$

$$\text{and} \quad \vartheta_C = \frac{\vartheta_{Cout} + \vartheta_{Cin}}{2} \tag{3.23}$$

in the equation for collector flow rate $\dot{m}'$ provides the *collector outlet temperature* ϑ_{Cout} for a given collector inlet temperature ϑ_{Cin}:

$$\vartheta_{Cout} = \frac{\eta_0 \cdot E + \dot{m}' \cdot c \cdot \vartheta_{Cin} + a \cdot (\vartheta_A - \frac{1}{2}\vartheta_{Cin})}{\dot{m}' \cdot c + \frac{1}{2}a} \tag{3.24}$$

If the collector flow rate is reduced to 18 kg/(m^2 h) and the collector inlet temperature ϑ_{Cin} is kept constant in the example above, the collector outlet temperature ϑ_{Cout} increases to 65°C and the average collector temperature ϑ_C increases to 50°C. As a result, the collector efficiency decreases from 70 to 65 per cent. These or even smaller flow rates are used in thermosyphon and so-called low-flow systems.

The collector flow rate may also be given in litre/h or litre/(m^2 h). This *volume flow* $\dot{V}$ depends on the *mass flow* $\dot{m}$ and the density ρ:

$$\dot{V} = \frac{1}{\rho} \cdot \dot{m} \tag{3.25}$$

For water with a density slightly lower than 1 kg/litre (or a density of about 1.06 kg/litre with added antifreeze agents), the numerical value of the volumetric flow is nearly equal to that of the mass flow.

The cross-sectional area A_P of the pipes in the collector cycle and the flow velocity v_P of the heat transfer medium defines the necessary *pipe diameter* d_P using $\dot{V} = A_P \cdot v_P = \pi \cdot \frac{1}{4} \cdot d_P^2 \cdot v_P$ as:

Table 3.8 *Parameters for Commercial Copper Pipes*

Name	Outside diameter in mm	Inside diameter in mm	Weight in kg/m	Capacity in l/m	Flow rate in l/h (for v_p=1 m/s)
12 × 1	12	10	0.31	0.079	280
15 × 1	15	13	0.39	0.133	480
18 × 1	18	16	0.51	0.201	720
22 × 1	22	20	0.59	0.314	1130
28 × 1.5	28	25	1.12	0.491	1770
35 × 1.5	35	32	1.41	0.804	2900
42 × 1.5	42	39	1.71	1.195	4300

Source: DIN, 1996

$$d_\mathrm{p} = \sqrt{\frac{4 \cdot \dot{m}}{\rho \cdot v_\mathrm{p} \cdot \pi}} = \sqrt{\frac{4 \cdot \dot{m}' \cdot A_\mathrm{C}}{\rho \cdot v_\mathrm{p} \cdot \pi}} \tag{3.26}$$

The necessary pipe diameter d_p for a system with a collector area of A_C = 5 m², a flow velocity of v_p = 1 m/s, an area-related mass flow of $\dot{m}'$ = 50 kg/(m²h)and a density of ρ = 1060 kg/m³ is a little less than 10 mm. Table 3.8 shows typical parameters of commercial copper pipes. A 12 × 1 copper pipe with an inner diameter of 10 mm would be suitable for such a system.

In practice, diameters a little larger are chosen. Friction inside the pipes slows the heat transfer medium and causes pressure losses, which can be reduced by choosing larger diameters. Tables 3.9 and 3.10 show *recommended piping lengths* for copper pipes in pumped and thermosyphon systems.

The total piping length is defined by the location of the collector and the storage tank in most cases. The following section calculates the heat losses for pipes with a chosen diameter and insulation. Therefore, a distinction is made between piping heat-up losses and circulation losses.

Table 3.9 *Recommended Diameters of Copper Pipes for Pumped Systems with Mixtures of Water and Antifreeze Agents*

| Collector area | Total piping length | | | | |
	10 m	20 m	30 m	40 m	50 m
below 5 m²	15 × 1	15 × 1	15 × 1	15 × 1	15 × 1
6–12 m²	18 × 1	18 × 1	18 × 1	18 × 1	22 × 1
13–16 m²	18 × 1	22 × 1	22 × 1	22 × 1	22 × 1
17–20 m²	22 × 1	22 × 1	22 × 1	22 × 1	22 × 1
21–25 m²	22 × 1	22 × 1	22 × 1	22 × 1	28 × 1.5
26–30 m²	22 × 1	22 × 1	28 × 1.5	28 × 1.5	28 × 1.5

Source: Wagner & Co, 1995

Table 3.10 *Recommended Diameters of Copper Pipes for Thermosyphon Systems with Mixtures of Water and Antifreeze Agents*

| | Height difference between collector and storage tank | | | | |
Collector area	0.5 m	1 m	2 m	4 m	6 m
below 4 m²	22 × 1	22 × 1	18 × 1	18 × 1	15 × 1
below 10 m²	28 × 1.5	28 × 1.5	22 × 1	22 × 1	18 × 1
below 20 m²	42 × 2	35 × 1.5	28 × 1.5	28 × 1.5	22 × 1

Source: Ladener, 1995

Piping heat-up losses

When the collector is not operational, for instance during the night, the connecting pipes and the fluid inside cool down until they reach the ambient temperature. When starting up the heating cycle again, pipes and heat transfer fluid must be heated up again first before they can transfer heat from the collector to the storage tank. Hence, energy is needed to heat both the pipes and the heat transfer fluid.

To heat pipes with mass m_P and heat capacity c_P as well as the heat transfer fluid with mass m_{HTF} and heat capacity c_{HFT} from temperature ϑ_1 to temperature ϑ_2, heat $Q_{Pheatup}$ is needed for n heat-up cycles:

$$Q_{Pheatup} = n \cdot (m_P \cdot c_P + m_{HTF} \cdot c_{HTF}) \cdot (\vartheta_2 - \vartheta_1) = n \cdot (m \cdot c)_{eff} \cdot (\vartheta_2 - \vartheta_1) \qquad (3.27)$$

The heat-up losses for a 20-m-long copper pipe with a cross-section of 15×1 mm are calculated as an example. It is assumed that the pipe is filled with a mixture of water and antifreeze agent and must be heated from a temperature of $\vartheta_1 = 20°C$ to $\vartheta_2 = 50°C$. The mass of the pipe with a heat capacity $c_P = 0.109$ Wh/(kg K) is $m_P = 7.8$ kg, the mass of the heat transfer fluid with heat capacity $c_{HTF} = 0.96$ Wh/(kg K) and density $\rho_{HTF} = 1.06$ kg/litre is $m_{HTF} = 2.82$ kg. With $(m \cdot c)_{eff} = 3.6$ Wh/K, the heat-up losses for one heat-up cycle ($n = 1$) come to:

$$Q_{Pheatup} = 108 \text{ Wh}$$

There are additional heat-up losses for heating the stop valves, pump and other components of the piping cycle. These losses can be calculated similarly if the mass and heat capacity of the components are known.

Piping circulation losses

Once the pipes and heat transfer medium are heated up, they still cause losses of heat to the environment. A pipe with length l, heat transition coefficient k' and with an ambient temperature of ϑ_A with continuous circulation over the circulation time t_{circ} produces *circulation losses* of:

$$Q_{\text{circ}} = k' \cdot l \cdot t_{\text{circ}} \cdot \left(\vartheta_{\text{HTF}} - \vartheta_{\text{A}} \right) \tag{3.28}$$

The *heat transition coefficient*

$$k' = \frac{\pi}{\dfrac{1}{2\lambda} \ln \dfrac{d_1}{d_p} + \dfrac{1}{\alpha \cdot d_1}} \tag{3.29}$$

can be calculated using the thermal conductivity λ of the insulation of a pipe of outer diameter d_p with the outer diameter of the insulation d_1. Heat conductivities λ of various materials are given in Table 3.3. For the *surface coefficient of heat transfer* α from the insulation to air, linearly interpolated values are used between $\alpha = 10$ W/(m² K) for $k' = 0.2$ W/(m K) and $\alpha = 15.5$ W/(m² K) for $k' = 0.5$ W/(m K).

The example of the previous section is also used here to calculate the circulation losses of a 20-m-long 15 × 1-mm copper pipe ($d_p = 15$ mm). The surface coefficient of heat transfer is taken to be $\alpha = 10$ W/(m² K). The heat transition coefficient for an insulation thickness of 30 mm ($d_1 = 0.075$ m) and a thermal conductivity of $\lambda = 0.040$ W/(m K) can be estimated from Equation (3.29) as $k' = 0.1465$ W/(m K). Hence, the piping circulation losses at an ambient temperature of $\vartheta_{\text{A}} = 20°$C for a heat transfer fluid temperature of $\vartheta_{\text{HTF}} = 50°$C for a circulation time of $t_{\text{circ}} = 8$ h become:

$$Q_{\text{circ}} = 703 \text{ Wh.}$$

The circulation losses of a 20-m-long pipe with a diameter of 22 mm and an insulation thickness of 10 mm for a temperature difference between heat transfer fluid and ambient air of 40°C and a circulation time of 10 h are as high as $Q_{\text{circ}} = 2500$ Wh. This demonstrates clearly that the insulation should be as good as possible to avoid loosing the majority of the heat on the way to the storage tank. Also, pipe lengths should be kept as short as possible and pipe clips must be attached without forming heat bridges. When laying pipes, it must be considered that they expand as a result of temperature changes.

If the regulator stops the circulation in the collector cycle, the pipes and the heat transfer medium cool down again. At a time t_1 with an ambient temperature of ϑ_{A} and a heat transfer fluid temperature of $\vartheta_{\text{HTF}}(t_1)$, the stored heat in the pipes is:

$$Q(t_1) = (c \cdot m)_{\text{eff}} \cdot \left(\vartheta_{\text{HTF}}(t_1) - \vartheta_{\text{A}} \right) \tag{3.30}$$

This heat is reduced by the heat flow:

$$\dot{Q} = k' \cdot l \cdot \left(\vartheta_{\text{HTF}}(t) - \vartheta_{\text{A}} \right) \tag{3.31}$$

The *stored heat* at time t_2 becomes:

$$Q(t_2) = Q(t_1) - k' \cdot l \cdot (\vartheta_{\text{HTF}}(t_1) - \vartheta_A) \cdot (t_2 - t_1) \tag{3.32}$$

with the resulting heat transfer fluid temperature:

$$\vartheta_{\text{HTF}}(t_2) = \frac{Q(t_2)}{(c \cdot m)_{\text{eff}}} + \vartheta_A = \left(1 - \frac{k' \cdot l \cdot (t_2 - t_1)}{(c \cdot m)_{\text{eff}}}\right) \cdot (\vartheta_{\text{HTF}}(t_1) - \vartheta_A) + \vartheta_A \tag{3.33}$$

However, this equation is only valid for small time intervals because the continuously changing temperature of the heat transfer fluid must be considered for the calculation of the heat flow. For longer time intervals, the temperature of the heat transfer fluid $\vartheta_{\text{HTF}}(t_2)$ at time t_2 can be calculated with the time interval $t_2 - t_1$ subdivided into n small time intervals Δt. With

$$\Delta t = \frac{t_2 - t_1}{n} \tag{3.34}$$

the temperature of the heat transfer fluid becomes:

$$\vartheta_{\text{HTF}}(t_2) = \left(1 - \frac{k' \cdot l \cdot \Delta t}{(c \cdot m)_{\text{eff}}}\right)^n \cdot (\vartheta_{\text{HTF}}(t_1) - \vartheta_A) + \vartheta_A \tag{3.35}$$

Finally, $\Delta t \to 0$, $t_1 = 0$ and $t_2 = t$ gives:

$$\vartheta_{\text{HTF}}(t) = \exp\left(-\frac{k' \cdot l}{(c \cdot m)_{\text{eff}}} \cdot t\right) \cdot (\vartheta_{\text{HTF0}} - \vartheta_A) + \vartheta_A \tag{3.36}$$

A 20-m-long 15×1 copper pipe with an insulation thickness of 30 mm (k' = 0.15 W/(m K)) and with $(c \cdot m)_{\text{eff}}$ = 3.6 Wh/K at an ambient temperature ϑ_A = 20°C cools down from $\vartheta_{\text{HTF}}(t_1)$ = 50°C to $\vartheta_{\text{HTF}}(t_2)$ = 33°C in 1 hour. If the collector cycle starts up again the pipes must be heated up again with the above-described heat-up losses.

THERMAL STORAGE

Different storage systems are used for solar thermal systems depending on the application. The objective of heat storage systems is to provide desired heat even at locations with fluctuating solar irradiances. They can be divided into two groups:

- short-term storage systems (daily cycle)
- long-term storage (inter-seasonal storage) systems.

Long-term storage systems should compensate for seasonal fluctuations, whereas short-term storage systems generally are only required to store heat

Table 3.11 *Parameters of Low-temperature Storage Materials*

	Density ρ in kg/m^3	Heat conductivity λ in W/(m K)	Heat capacity c in kJ/(kg K)
Water (0°C)	999.8	0.5620	4.217
Water (20°C)	998.3	0.5996	4.182
Water (50°C)	988.1	0.6405	4.181
Water (100°C)	958.1	0.6803	4.215
Granite	2750	2.9	0.890
Coarse gravely soil	2040	0.59	1.840
Clay soil	1450	1.28	0.880
Concrete	2400	2.1	1.000

Source: Khartchenko, 1998

over periods of a few hours to a few days. Hence, long-term storage systems need significantly larger volumes. Large storage systems can be:

- artificial storage basins
- rock caverns (cavities in rocks)
- aquifer storage (groundwater storage)
- ground and rock storage.

Furthermore, storage systems can be divided into different temperature ranges:

- low-temperature storage systems for temperatures below 100°C
- medium-temperature storage systems for temperatures between 100°C and 500°C
- high-temperature storage systems for temperatures above 500°C.

Finally, there are different types of heat storage such as:

- storage of sensible (noticeable) heat
- storage of latent heat (storage due to changes in physical state)
- thermo-chemical energy storage.

Because of space considerations, this book describes only *low-temperature storage systems for the storage of sensible heat*. Table 3.11 shows parameters of some low-temperature storage materials.

Domestic hot water storage tanks

Only annual usage simulations can estimate the exact dimensions of a hot water storage tank. The size depends on the hot water demand, solar fraction (see Equation 3.59, p112), collector performance, collector orientation, pipes and last but not least on the annual solar irradiation. For central European climates, a rough estimate can be made. Here, the *storage volume* should be

1.5–2 times the daily demand. Besides the storage volume for the daily demand, a standby volume of 50 per cent and a preheating volume of 20 litres per square metre of collector surface should be considered. Commercial pressurized hot water tanks are available in sizes from less than 100 litres to more than 1000 litres. The recommended storage size for a one-family house with four to six persons is between 300 litres and 500 litres.

Most solar storage tanks have two heat exchangers (see also Figure 3.4). The heat exchanger of the solar cycle is in the lower part of the storage tank and the heat exchanger for the auxiliary heater is in the upper part. The tank has an opening near the middle of the heat exchangers for integrated temperature sensors for the control system. The cold water inlet is at the bottom and the hot water outlet at the top of the storage tank to achieve good heat stratification.

Figure 3.15 shows a horizontal, cylindrical hot water storage tank with spherical ends. Storage losses are calculated for this tank, as an example. Heat storage tanks always suffer *heat losses* due to heat transition through the insulation. Good insulation should have a thickness of at least 100 mm for a heat conductivity of λ = 0.04 W/(m K). Some new materials have very low heat conductivities; for instance, a super-insulation glass fibre vacuum insulation can reach heat conductivities of λ = 0.005 W/(m K) at pressures below 10–3 mbar.

The heat storage capacity of a hot water tank is:

$$Q = c \cdot m \cdot (\vartheta_S - \vartheta_A)$$
(3.37)

This heat capacity depends on the temperature difference between the average storage temperature ϑ_S and the ambient temperature ϑ_A as well as on the heat capacity c and mass m of the storage medium. The heat capacity of water at a temperature of 50°C and with a density of ρ_{H_2O} = 0.9881 kg/litre is c_{H_2O} = 4.181 kJ/(kg K) = 1.161 Wh/(kg K). Hence, the heat storage capacity of a 300-litre hot water storage tank with a temperature difference of 70°C is 24 kWh.

The *storage losses* $\dot{Q}_S$ of a cylindrical and spherical hot water storage tank are the sum of the losses $\dot{Q}_{S,cyl}$ of the cylindrical part and the losses $\dot{Q}_{S,sphere}$ of both spherical caps:

$$\dot{Q}_S = \dot{Q}_{S,cyl} + 2 \cdot \dot{Q}_{S,sphere}$$
(3.38)

The losses in the cylindrical part

$$\dot{Q}_{S,cyl} = k' \cdot l_{cyl} \cdot (\vartheta_S - \vartheta_A)$$
(3.39)

can be calculated similarly to the losses of the pipes in the previous sections with the heat transition coefficient k', the length l_{cyl} and the difference between the average storage temperature ϑ_S and the ambient temperature ϑ_A.

The heat conductivity of the insulation λ, the surface coefficient of heat transfer α between insulation and air as well as the outer diameter d_o and the

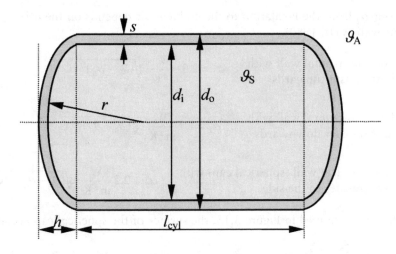

Figure 3.15 *Cylindrical Hot Water Tank with Spherical Ends*

inner diameter d_i of the heat insulation of the cylindrical part define the heat transition coefficient:

$$k' = \frac{\pi}{\frac{1}{2 \cdot \lambda} \cdot \ln \frac{d_o}{d_i} + \frac{1}{\alpha \cdot d_o}}$$

(3.40)

Linearly interpolated values between 10 W/(m^2 K) for k' = 0.2 W/(m K) and 15.5 W/(m^2 K) for k' = 0.5 W/(m K) are used to estimate the surface coefficient of heat transfer α.

With the temperature difference between the storage medium and ambient air, the coefficient of heat transfer k and the surface A_{sphere} of the spherical caps, the heat losses of the spherical caps become:

$$\dot{Q}_{S,sphere} = k \cdot A_{sphere} \cdot (\vartheta_S - \vartheta_A)$$

(3.41)

With the surface coefficient of heat transfer α_1 from the tank wall to the insulation and α_2 from the insulation to the ambient air, the insulation thickness s, the heat conductivity λ of the insulation, and assuming that the temperature of the tank wall is equal to the storage temperature ϑ_S, the *coefficient of heat transfer* becomes:

$$k = \frac{1}{\frac{1}{\alpha_1} + \frac{1}{\alpha_2} + \frac{s}{\lambda}}$$

(3.42)

The surface coefficient of heat transfer from the tank wall to the insulation can be estimated as α_1 = 300 W/(m^2 K). The surface coefficient of heat transfer α_2 from the insulation to the ambient air depends on the orientation of the wall (VDI, 1982).

For a horizontal wall with heat transfer upwards:

$$\alpha_2 = 2.3 \frac{W}{m^2 K} \cdot \sqrt[4]{(\vartheta_S - \vartheta_A)/\,°C}$$

For a horizontal wall with heat transfer downwards:

$$\alpha_2 = 1.7 \frac{W}{m^2 K} \cdot \sqrt[4]{(\vartheta_S - \vartheta_A)/\,°C}$$

For a vertical wall (spherical cap) with heat transfer to the side:

$$\alpha_2 = 2.2 \frac{W}{m^2 K} \cdot \sqrt[4]{(\vartheta_S - \vartheta_A)/\,°C}$$

With r and h defined in Figure 3.15, the surface of the spherical cap becomes:

$$A_{sphere} = 2\pi \cdot r \cdot h \tag{3.43}$$

The following example calculates the heat losses of a 300-litre storage tank. The ambient temperature and storage temperature are ϑ_A = 20°C and ϑ_S = 90°C, the dimensions of the tank are l_{cyl} = 1.825 m, d_o = 0.7 m, d_i = 0.5 m, r = 0.45 m, h = 0.11 m and s = 0.1 m. With the heat conductivity of the insulation λ = 0.035 W/(m K) and the surface coefficient of heat transfer α = 15.5 W/(m² K) in the cylindrical part, the heat transition coefficient is k' = 0.64 W/(m K). With the surface area of the spherical cap of A_{sphere} = 0.311 m² and the heat transition coefficient of the spherical cap k = 0.33 W/(m² K), the storage losses become:

$$\dot{Q}_S = (k' \cdot l_{cyl} + 2 \cdot k \cdot A_{sphere}) \cdot (\vartheta_S - \vartheta_A) = 96 \text{ W}$$

The temperature ϑ_S of a stationary storage tank decreases with the time t. Thus, the storage losses decrease as well. If no heat is fed into or taken from the storage tank, the storage temperature

$$\vartheta_S(t) = \exp\left(-\frac{k' \cdot l_{cyl} + 2 \cdot k \cdot A_{sphere}}{c \cdot m} \cdot t\right) \cdot (\vartheta_{S0} - \vartheta_A) + \vartheta_A \tag{3.44}$$

can be calculated as described for the pipes above. The value

$$\tau = \frac{c \cdot m}{k' \cdot l_{cyl} + 2 \cdot k \cdot A_{sphere}} \tag{3.45}$$

is the *time constant* of the storage. It describes the time taken for the temperature difference to decrease to $1/e$ = 36.8 per cent of the initial value. The time constant of the example above is τ = 250 h = 10.4 days.

Figure 3.16 shows the storage temperature of a stationary storage tank. It is obvious that a high portion of the stored heat is emitted again to the

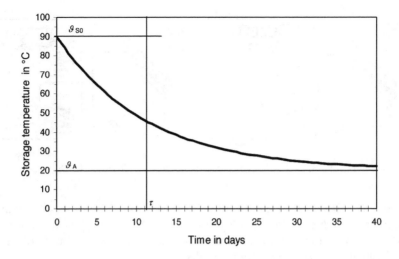

Figure 3.16 *Storage Temperature* ϑ_S *for a 300-litre Storage Tank without Loading or Unloading*

environment. After more than 1 week, the storage temperature is only half the initial value. Such a storage tank can therefore only maintain the temperature for a few days. The ratio of volume to surface area increases for larger storage tanks, so the relative heat losses decrease. Storage systems of 1000 m^3 or more can achieve time constants of about 6 months. Such systems can be used for seasonal storage, i.e. heat storage from summer to winter. Good insulation is also very important for seasonal storage systems to keep the storage losses low.

However, realistic integrated solar thermal system hot water storage is usually not operated under stationary conditions. The solar collector feeds heat continuously into the storage tank and consumers take hot water from the tank. Then, only detailed computer programs can estimate the variation of the storage temperature (see CD-ROM).

The storage temperatures calculated above are all average temperatures; however, most storage tanks have desirable *temperature stratifications*. The temperature at the top of the storage tank, i.e. near the water outlet, is higher than at the bottom, near the water inlet. This stratification can be considered if the storage is subdivided into several layers. In this case, the heat flow between all layers must be calculated separately.

Large systems for higher demands, such as for family houses, often use *two storage tanks*. Here, the tanks are connected in series. For a system with solar preheating, the first storage tank contains the solar heat exchanger and the second storage tank contains the heat exchanger for the auxiliary heater. The solar fraction of these systems is relatively low and the profitability of this system compared to one with low conventional fuel prices is relatively high. Another concept with two storage tanks contains a solar heat exchanger in both tanks. The solar collector can heat up both tanks separately. This can

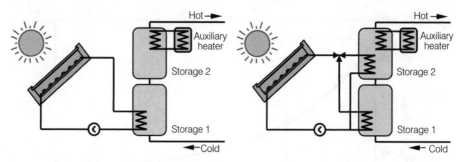

Figure 3.17 *Collector Systems with Two Storage Tanks*

increase the solar fraction significantly. If the first storage tank is fully loaded, the control switches to the second storage tank. An auxiliary heater can heat up the second storage tank in both systems. Figure 3.17 shows the two described concepts with two storage tanks.

Swimming pools

The pool itself is the heat storage medium of a swimming pool system. The usefulness of the system lies in the storage. The solar heating system must compensate only for the storage losses, i.e. the losses of the pool (Figure 3.18). The pool losses $\dot{Q}_{Pool,loss}$ consist of convection losses $\dot{Q}_{conv}$, radiation losses $\dot{Q}_{rad}$, vaporization losses $\dot{Q}_{vap}$ and transmission losses $\dot{Q}_{trans}$:

$$\dot{Q}_{Pool,loss} = \dot{Q}_{conv} + \dot{Q}_{rad} + \dot{Q}_{vap} + \dot{Q}_{trans} \tag{3.46}$$

The transmission losses $\dot{Q}_{trans}$ from the swimming pool to the earth can be neglected for solar-heated outdoor pools during the summer season.

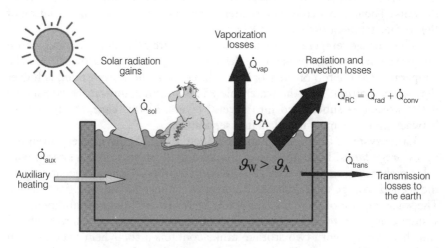

Figure 3.18 *Energy Balance of a Swimming Pool*

With the ambient air temperature ϑ_A, the water temperature ϑ_W, the water surface area A_W and the heat transfer coefficient given by:

$$\alpha_{conv} = \left(3.1 + 4.1 \cdot v_{wind} \cdot \frac{s}{m} \right) \cdot \frac{W}{m^2 K} \tag{3.47}$$

the convection losses become:

$$\dot{Q}_{conv} = \alpha_{conv} \cdot A_W \cdot (\vartheta_W - \vartheta_A) \tag{3.48}$$

The wind speed v_{wind} at a height of 0.3 m above the water surface can be calculated from measurements at other heights using the equations in Chapter 5, section headed 'Influence of surroundings and height', p185.

Radiation exchange between the swimming pool surface and the sky causes *radiation losses*. With the Stefan–Boltzmann constant $\sigma = 5.67051 \cdot 10^{-8}$ W/(m² K⁴), the emittance ε_W of water ($\varepsilon_W \approx 0.9$), water surface area A_W and absolute temperatures T_W and T_{Sky} of the water and the sky, respectively, the radiation losses become:

$$\dot{Q}_{rad} = \sigma \cdot \varepsilon_W \cdot A_W \cdot \left(T_W^4 - T_{Sky}^4 \right) \tag{3.49}$$

The *sky temperature*

$$T_{Sky} = T_A \cdot \left(0.8 + \frac{\vartheta_{dew}(\vartheta_A)}{250°C} \right)^{0.25} \tag{3.50}$$

can be calculated using the absolute ambient air temperature T_A in K and the dew-point temperature ϑ_{dew} (Smith et al, 1994).

The *dew-point temperature*

$$\vartheta_{dew} = 234.175\ °C \cdot \frac{\ln\left(\dfrac{\varphi \cdot p}{0.61078\ kPa} \right)}{17.08085 - \ln\left(\dfrac{\varphi \cdot p}{0.61078\ kPa} \right)} \tag{3.51}$$

can be estimated from the humidity φ and the *saturated vapour pressure p*:

$$p = 0.61078\ kPa \cdot \exp\left(\frac{17.08085 \cdot \vartheta_A}{234.175\ °C + \vartheta_A} \right) \tag{3.52}$$

The saturated vapour pressure p depends on the ambient air temperature ϑ_A and is measured in Pascals (1 Pa = 1 N/m² = 0.01 mbar). The mean relative humidity φ in moderate climates (e.g. Germany) is about 70 per cent during the outdoor pool season.

Table 3.12 *Saturated Vapour Pressure p of Water and the Dew-point Temperature ϑ_{dew} at 70 per cent Relative Air Humidity as a Function of the Ambient Air Temperature ϑ_A*

ϑ_A in °C	10	12	14	16	18	20	22	24	26	28	30
p in kPa	1.23	1.40	1.60	1.82	2.07	2.34	2.65	2.99	3.37	3.79	4.25
ϑ_{dew} in °C	4.8	6.7	8.6	10.5	12.5	14.4	16.3	18.2	20.1	22.0	23.9

Table 3.12 shows the saturated vapour pressures and the dew-point temperatures calculated using equations (3.51) and (3.52) for various temperatures.

With the vaporizing mass flow $\dot{m}_V$ (eg kg of water/hour) and the heat of vaporization h_V = 2.257 kJ/kg of water, the vaporization losses are found from:

$$\dot{Q}_{vap} = h_V \cdot \dot{m}_V \tag{3.53}$$

However, empirical equations are often used. When calculating the vaporization losses using the wind speed v_{Wind} at height 0.3 m and using the saturated vapour pressure p, the water temperature ϑ_W, the ambient air temperature ϑ_A, the relative humidity φ and the swimming pool surface area A_W, we get (Hahne and Kübler, 1994):

$$\dot{Q}_{vap} = A_W \cdot (0.085\frac{m}{s} + 0.0508 \cdot v_{Wind}) \cdot (p(\vartheta_W) - \varphi \cdot p(\vartheta_A)) \tag{3.54}$$

Neglecting the transmission losses, the total losses from a swimming pool with a surface area of A_W = 20 m², a wind speed v_{Wind} = 1 m/s, ambient air temperature ϑ_A = 20°C, water temperature ϑ_W = 24°C and relative humidity φ = 0.7 become:

$$\dot{Q}_{Pool,loss} = \dot{Q}_{conv} + \dot{Q}_{rad} + \dot{Q}_{vap} = 576\ W + 1493\ W + 3672\ W = 5741\ W$$

A large amount of energy would thus be needed to compensate for the losses. However, solar radiation onto the pool surface produces gains that reduce the losses significantly.

With solar irradiance E, water surface area A_W and absorptance α_{abs}, the solar radiation gains $\dot{Q}_{sol}$ are estimated as:

$$\dot{Q}_{sol} = \alpha_{abs} \cdot E \cdot A_W \tag{3.55}$$

The absorptance α_{abs} of pools with white tiles is about 0.8, with light blue tiles 0.9 and with dark blue tiles over 0.95. The absorptance also increases with the depth of the water. A solar irradiance of E = 319 W/m² at an absorptance of 0.9 already compensates for the losses of the 20 m² swimming pool from the example above.

Covering the swimming pool during the night can reduce the heat losses by about 40–50 per cent.

The heat demand

$$Q_H = \sum_{i=0}^{t_{tot}/\Delta t} \left(\dot{Q}_{Pool,loss}(i \cdot \Delta t) - \dot{Q}_{sol}(i \cdot \Delta t)\right)\Delta t \stackrel{\Delta t \to 0}{=} \int_{t=0}^{t_{tot}} \left(\dot{Q}_{Pool,loss}(t) - \dot{Q}_{sol}(t)\right)\mathrm{d}t \quad (3.56)$$

of a swimming pool can then be estimated by considering the difference between the losses and the solar radiation gains over the operating period t_{tot}.

In moderate climatic regions (e.g. central Europe), the heat demand of a swimming pool without a cover and at a base temperature of 23°C is about 300 kWh/m² per season. This demand can be provided easily by a solar heating system. The solar absorber size should be 50–80 per cent of the pool surface; however, these values can vary significantly. Therefore, simulations with professional computer programs can provide more exact values.

HEAT DEMAND AND SOLAR FRACTION

The heat demand Q_D of domestic water systems can be calculated from the amount of water taken. With the heat capacity of water [c_{H_2O} = 1.163 Wh/(kg K)], the taken water mass m, the cold water temperature ϑ_{CW} and the warm water temperature ϑ_{HW}, the heat demand becomes:

$$Q_D = c \cdot m \cdot (\vartheta_{HW} - \vartheta_{CW}) \quad (3.57)$$

Table 3.13 shows typical values for the *hot water demand* in residential buildings in Germany. If no value for the cold water temperature is given, a value of ϑ_{CW} = 10°C can be used. In countries with higher annual ambient temperatures, a higher cold water temperature should be chosen. If washing machines or dishwashers with hot water inlets are used, the hot water demand increases.

Table 3.14 shows typical values for the *hot water demand of hotels*, hostels and pensions. The hot water demand of restaurants can be estimated as 230–460 Wh/ per set menu and of saunas as 2500–5000 Wh/user.

Table 3.13 *Hot Water Demand of Residential Buildings in Germany*

	Hot water demand in litres/(day and person)		Specific heat content in Wh/(day and person)
	ϑ_{HW} = 60°C	ϑ_{HW} = 45°C	
Low demand	10–20	15–30	600–1200
Average demand	20–40	30–60	1200–2400
High demand	40–80	60–120	2400–4800

Source: VDI, 1982

Table 3.14 *Hot Water Demand of Hotels, Hostels and Pensions in Germany*

	Hot water demand in litres/(day and person)		Specific heat content in Wh/(day and person)
	$\vartheta_{HW} = 60°C$	$\vartheta_{HW} = 45°C$	
Room with bath	95–138	135–196	5500–8000
Room with shower	50–95	74–135	3000–5500
Other rooms	25–35	37–49	1500–2000
Hostels and pensions	25–50	37–74	1500–3000

Source: VDI, 1982

Because the tables give only a rough estimate, it is recommended that a more exact analysis be made when planning a solar thermal system. Table 3.15 gives some further information.

Besides the annual hot water requirements, the demand on shorter timescales must also be considered. If there are significant differences between various days or considerable seasonal differences between summer and winter, the sizing of the solar energy system changes. In this case, computer simulations are prerequisites for an exact system sizing and output prediction.

The heat demand Q_D, the heat losses (piping heat-up losses $Q_{Pheatup}$, circulation losses Q_{circ} in the collector cycle and storage losses Q_S) and the solar collector power output Q_{out} define the amount of auxiliary heat required:

$$Q_{aux} = Q_D + Q_{Pheatup} + Q_{circ} + Q_S - Q_{out} \tag{3.58}$$

Besides piping losses in the collector cycle, there are also piping losses from the storage to the taps. These losses can be calculated similarly to the losses in the collector cycle. Simulations usually calculate the heat flows and add them to give annual heat values.

An important parameter for solar energy systems is the *solar fraction SF*. This parameter describes the share of the heat demand provided by the solar thermal system. It is defined as the ratio of the heat that the solar cycle feeds into the storage to the total heat demand that consists of the heat demand Q_D and the storage losses Q_S:

$$SF = \frac{Q_D + Q_S - Q_{aux}}{Q_D + Q_S} = \frac{Q_{out} - Q_{Pheatup} - Q_{circ}}{Q_D + Q_S} \tag{3.59}$$

Table 3.15 *Hot Water Usage for Various Activities*

	Demand	Temperature	Heat content
Dishwashing per person	12–15 l/day	50°C	550–700 Wh/day
Hand wash (one time)	3–5 l	37°C	95–160 Wh
Bath (one time)	150 l	40°C	5200 Wh
Shower (one time)	30–45 l	37°C	940–1400 Wh
Hair wash (one time)	10–15 l	37°C	310–470 Wh

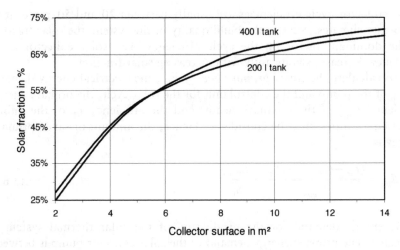

Note: Location: Berlin, Collector Inclination: 30°, Heat Demand: 10 kWh/day

Figure 3.19 *Solar Fraction as a Function of the Collector Surface: Simulation by the Software Getsolar.*

Solar energy systems for domestic water heating in moderate climates are usually designed for solar fractions of about 50–60 per cent. This is a compromise between desired high solar fractions and economic considerations. However, the solar fraction in regions with high annual irradiations can be much higher. If there are only small differences between summer and winter, the solar fraction can approach 100 per cent.

Figure 3.19 shows the solar fraction as a function of the collector surface for two different volumes of storage tank calculated for Berlin. The heat demand was constant for all calculations. The curve shows that the solar fraction increases very quickly with the collector surface area for relatively small collectors. However, in the given example, the collector surface must be doubled to increase the solar fraction from 60 to 70 per cent. Large storage volumes are counter-productive for small collector surfaces because the storage losses are disproportionately high in such a case. If the solar collector surface increases significantly, the storage volume should be increased as well.

Another important parameter for analysis is the *solar collector cycle efficiency* η_{CC}. It describes the total efficiency of the solar thermal system. It is defined as the ratio of the annual heat that the solar cycle feeds into the storage system to the annual solar radiant energy on the collector surface. With annual irradiation H_{Solar} on the collector surface and collector area A_C, the collector cycle efficiency becomes:

$$\eta_{CC} = \frac{SF \cdot (Q_D + Q_S)}{H_{Solar} \cdot A_C} = \frac{Q_{out} - Q_{Pheatup} - Q_{circ}}{H_{Solar} \cdot A_C} \qquad (3.60)$$

Solar collector cycle efficiencies are usually between 20 and 50 per cent in temperate climates. Next to the build quality of the system, the solar fraction is the dominant factor on the cycle efficiency. The solar collector cycle efficiency decreases significantly with increasing solar fraction.

To calculate the *saved primary energy* E_{PE}, the electrical energy demand E_{pump} of the pump and the control unit for the solar cycle, the primary energy efficiency η_{aux} of the auxiliary heater and the efficiency η_E of the public electricity utilities must be considered. Hence, the amount of saved primary energy is:

$$E_{PE} = \frac{Q_{out} - Q_{Pheatup} - Q_{circ}}{\eta_{aux}} - \frac{E_{Pump}}{\eta_E} \qquad (3.61)$$

if the energy demand for the production of the solar thermal system is neglected. The primary energy demand of the solar collector pump is between 2 and 15 per cent of the collector output. To keep the pumping energy demand low, the diameters of the pipes and the design of the pump must be estimated carefully. The simulation programs on the CD-ROM accompanying this book can provide further advice for this task.

Chapter 4

Photovoltaics

INTRODUCTION

The word 'photovoltaic' consists of the two words, photo and Volta. Photo stands for light (Greek phõs, photós: light) and Volta (Count Volta, 1745–1827, Italian physicist) is the unit of the electrical voltage. In other words, photovoltaic means the direct conversion of sunlight to electricity. The common abbreviation for photovoltaic is PV.

The history of photovoltaics goes back to the year 1839, when Becquerel discovered the photo effect, but in that century the technology was not available to exploit this discovery. The semiconductor age began about 100 years later. After Shockley had developed a model for the p–n junction, Bell Laboratories produced the first solar cell in 1954. The efficiency of this cell was about 5 per cent. Initially, cost was not a major issue, because the first cells were designed for space applications.

In the following years, solar cell efficiency increased continuously; laboratory silicon solar cells have reached efficiencies of around 25 per cent today. The main material used in the construction of solar cells is still silicon, but other materials have been developed, either for their potential for cost reduction or their potential for high efficiency. Costs have decreased significantly in recent decades; nevertheless, photovoltaic electricity generating costs are still higher than the costs of conventional power plants (see Chapter 6). Due to high growth rates in the photovoltaic sector, cost reduction will continue.

Photovoltaics offer the highest versatility among renewable energy technologies. One advantage is the *modularity*. All desired generator sizes can be realized, from the milliwatt range for the supply of wristwatches or pocket calculators to the megawatt range for the public electricity supply.

Many photovoltaic applications are built into consumer appliances or relate to leisure activities or off-grid site supply, for example, telecommunications or solar home systems. In several countries, particularly in Japan and Germany, large governmental programmes were initiated, advancing grid-connected installations. Tens of thousands of grid-connected systems that have been installed since the early 1990s have proven the suitability of the technology. The potential for photovoltaic installations is enormous. Theoretically, PV systems could cover the whole electricity demand of most countries in the world.

Figure 4.1 *Roof-integrated Photovoltaic System [Georgetown University in Washington (USA), Installed in 1984 with 4464 Photovoltaic Modules and 337 kW$_p$ Total Power]*

Figure 4.1 shows an example of a *roof-integrated photovoltaic system*, which was installed in the 1980s in the US. Various megawatt roof-integrated PV systems are installed today. One example is the 1-MW photovoltaic system at the Munich exhibition halls, which was finished in 1997. Altogether 7812 photovoltaic modules on a roof area of 38,100 m^2 generate about one million kWh of electricity per year. This equals the average consumption of 340 German households.

Worldwide, the installed photovoltaic capacity and the share of electricity generated by PV are still low, despite impressive market growth. The political environment and magnitude of market introduction programmes will determine the future of this technology.

This chapter describes the fundamentals of PV systems such as the operating principles, calculations of the energy yield and systems and their applications. Since the descriptions use many electrical quantities, Table 4.1 gives an overview of them.

OPERATION OF SOLAR CELLS

Bohr's atomic model

This section describes the operating principles of semiconductor solar cells; therefore, an understanding is necessary of what a semiconductor is. The section starts with Bohr's atomic model for single atoms and continues with the energy band model for solids.

Table 4.1 *Overview of the Most Important Electrical Quantities*

Name	Symbol	Unit	Name	Symbol	Unit
Electrical energy	W, E	W s, J	Specific resistance	ρ	Ω m
Electrical power	$P=V\cdot I$	W	Electrical field strength	$E=-dV/ds$	V/m
Voltage	V	V	Inductivity (inductance)	L	H=Vs/A
Current	I	A	Capacity	C	F=As/V
Resistance	$R=V/I$	Ω	Electrical charge	$Q=\int I dt$	C=As
Conductance	$G=1/R$	℧	Force in electrical field	$F=E\cdot Q$	N

According to Bohr's atomic model, electrons with rest mass:

$$m_e = 9.1093897 \cdot 10^{-31} \text{ kg} \tag{4.1}$$

revolve around the atomic nucleus in an orbit with radius r_n and angular frequency ω_n. This orbital movement results in a *centrifugal force*:

$$F_Z = m_e \cdot r_n \cdot \omega_n^2 \tag{4.2}$$

Electrons, each with the *elementary charge of an electron*

$$e = 1.60217733 \cdot 10^{-19} \text{ A s} \tag{4.3}$$

are held in obit around the nucleus of an atom (which consists of Z positively charged protons and additional uncharged neutrons) by the attractive *Coulomb force*:

$$F_C = \frac{1}{4\pi \cdot \varepsilon_0} \cdot \frac{Z \cdot e^2}{r_n^2} \tag{4.4}$$

where $\varepsilon_0 = 8.85418781762 \cdot 10^{-12} \dfrac{\text{A s}}{\text{V m}}$ \hfill (4.5)

and is called the *permittivity* or dielectric constant. The Coulomb and the centrifugal force are balanced, keeping the electron in its orbit. According to Planck's theorem, electrons can only remain in orbits where the orbital angular momentum is an integer multiple of:

$$\hbar = \frac{h}{2 \cdot \pi} \tag{4.6}$$

which is derived from **Planck's constant**:

$$h = 6.6260755 \cdot 10^{-34} \text{ J s} \tag{4.7}$$

The quantization of the orbital angular momentum leads to:

$$n \cdot \hbar = m_e \cdot r_n^2 \cdot \omega_n \qquad (4.8)$$

With this expression, the balance of forces $F_Z = F_C$ can be resolved for the orbital radius, which becomes:

$$r_n = \frac{n^2 \cdot \hbar^2 \cdot 4 \cdot \pi \cdot \varepsilon_0}{Z \cdot e^2 \cdot m_e} \qquad (4.9)$$

and the *angular frequency*:

$$\omega_n = \frac{1}{(4 \cdot \pi \cdot \varepsilon_0)^2} \cdot \frac{Z^2 \cdot e^4 \cdot m_e}{\hbar^3 \cdot n^3} \qquad (4.10)$$

The integer index n describes the orbit number. An electron in orbit n contains the energy:

$$E_n = \tfrac{1}{2} \cdot m_e \cdot v_e^2 = \tfrac{1}{2} \cdot m_e \cdot r_n^2 \cdot \omega_n^2 = \frac{1}{n^2} \cdot \frac{Z^2 \cdot e^4 \cdot m_e}{32 \cdot \pi^2 \cdot \varepsilon_0^2 \cdot \hbar^2} \qquad (4.11)$$

For instance, the energy of an electron in a hydrogen atom with the proton number $Z = 1$ at the first orbit ($n = 1$) is $E_{1,Z=1} = 13.59$ eV.

Elevating an electron from one orbit to the next highest orbit requires the energy $\Delta E = E_n - E_{n+1}$. This energy must be provided from outside the atom. All orbits can only hold a limited number of electrons. At the first orbit ($n = 1$) the maximum number is 2 electrons, at the second 8, then 18, 32, 50 and so on. The electron energy decreases with rising orbit index n. For $n = \infty$ it becomes $E_\infty = 0$.

The photo effect

Light, with its photon energy, can provide the energy to lift an electron to a higher orbit. The *photon energy* is given by:

$$E = \frac{h \cdot c}{\lambda} \qquad (4.12)$$

with the wavelength λ and the speed of light $c = 2.99792458 \cdot 10^8$ m/s. When a photon with an energy of 13.59 eV hits a hydrogen atom electron in the first orbit, this energy is sufficient to lift the electron to orbit E_∞. In other words, it totally separates the electron from the nucleus. This energy is also called the ionization energy. The total release of an electron from the nucleus by photons is called the *external photoelectric effect*. The photon in the hydrogen example must have a wavelength lower than $\lambda = 90$ nm, which places it in the realm of X-rays.

Because photovoltaic cells mainly convert to electricity photons of visible, ultraviolet and infrared light, i.e. photons of lower energy than X-rays, the external photo effect is not applicable to photovoltaic cells. The so-called

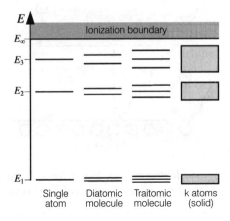

Figure 4.2 *Energy States of Electrons in Atoms, Molecules and Solids*

internal photo effect determines the effect of light in a solar cell, and is described in the following sections.

Whereas electrons in single atoms take clearly defined energy states, this is not the case in molecules with multiple electrons. Interactions between electrons of molecules with several atoms smear the energy states to narrow neighbouring levels. In a solid with k atoms, these levels are so close to each other that it is no longer possible to separate them. Here the single energy states of the electron orbits become so-called energy bands (see Figure 4.2). However, these energy bands can also carry only a limited number of electrons.

In the energy band model, electrons fill the bands one after another starting with the first, lowest energy band. The highest fully occupied band is called the *valence band* (VB). The next highest band, which can be partially occupied or totally empty, is called the *conduction band* (CB). The space between the valence band and conduction band contains forbidden energy states and is therefore called the *forbidden band* (FB). The energy gap between the bands is called the *band gap* E_g.

Solids are subdivided into conductors, semiconductors and isolators depending on the arrangement and occupation of the energy bands (see Figure 4.3). Electrons in *conductors* only fill the conduction band partially. The

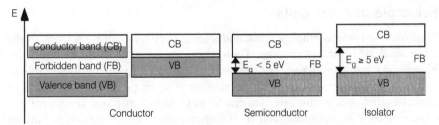

Figure 4.3 *Energy Bands of Conductors, Semiconductors and Isolators*

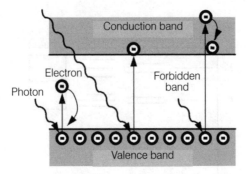

Figure 4.4 *The Lifting of Electrons from the Valence Band to the Conduction Band Caused by Light Energy in a Semiconductor (Inner Photo Effect)*

conduction band and the valence band can also overlap. Electrons can move within the solid and contribute to the electron conductivity in a partially filled conduction band. The specific electrical resistance of conductors is very low ($\rho < 10^{-5}$ Ω m). Most conductors are metallic materials.

The specific electrical resistance of isolators is high ($\rho > 10^7$ Ω m). The conduction band is totally empty and a relatively large amount of energy is needed to elevate electrons from the valence band to the conduction band due to the high band gap ($E_g \geqslant 5$ eV) in isolators.

Semiconductors have the most relevance for photovoltaics. The specific electrical resistance is between 10^{-5} Ω m and 10^7 Ω m. The conduction bands of semiconductors are empty, as in the case of isolators. However, due to the lower band gap ($E_g < 5$ eV), electrons can be more easily be lifted to the conduction band (Figure 4.4). The elevation of electrons into the conduction bands by photons is called the inner or *internal photo effect*.

An incident photon can have several effects: photon energy lower than the band gap will not elevate an electron because it cannot bridge the band gap. Photons with energy larger than the band gap can elevate electrons into the conduction band with a part of its energy. Surplus energy is lost, because the electron falls back to the edge of the conduction band.

Photoresistors, which change their resistance depending on the irradiance, use the internal photo effect. Photovoltaic cells also use the internal photo effect for generating current.

Principle of solar cells

Photovoltaic systems employ semiconductors. They have four electrons in the outer shell, or orbit, on average. These electrons are called valence electrons. Elementary semiconductors are elements of group IV of the periodic table of elements, for instance silicon (Si), germanium (Ge) or tin (Sn). Compounds of two elements containing one element from group III and one from group V (so-called III-V compounds) and II-VI compounds or combinations of various elements also have four valence electrons on average. An example of a III-V semiconductor is gallium arsenide (GaAs) and an example of a II-VI

Table 4.2 *Band Gap for Various Semiconductors at 300 K*

IV semiconductors		III-V semiconductors		II-VI semiconductors	
Material	E_g	Material	E_g	Material	E_g
Si	1.107 eV	GaAs	1.35 eV	CdTe	1.44 eV
Ge	0.67 eV	InSb	0.165 eV	ZnSe	2.58 eV
Sn	0.08 eV	InP	1.27 eV	ZnTe	2.26 eV
		GaP	2.24 eV	HgSe	0.30 eV

Source: data from Lechner, 1992

semiconductor is cadmium telluride (CdTe). Table 4.2 shows the different band gaps of various semiconductors.

Silicon is the material most commonly used in photovoltaics. Silicon is the second most abundant element in the earth's crust after oxygen, but it cannot be found in a chemically pure form. Silicon is an elementary semiconductor of the group IV of the periodic table of elements, i.e. silicon has four valence electrons in the outer shell. In order to obtain the most stable electron configuration, two electrons of neighbouring atoms in the silicon crystal lattice form an electron pair binding. In other words, two atoms jointly use these electrons. Electron pair bindings (covalent bonds) with four neighbours give silicon a stable electron configuration similar to that of the noble gas argon (Ar). In the energy band model, the valence band is now fully occupied and the conduction band is empty. Supplying sufficient energy by incident light or heat can elevate an electron from the valence band into the conduction band. This electron now can move freely through the crystal lattice. A so-called defect electron, or hole, remains in the valence band. Figure 4.5 illustrates this process. The formation of defect electrons is responsible for the so-called *intrinsic conduction* of semiconductors.

Electrons and holes always arise in pairs, i.e. there are exactly as many electrons as holes. This is described by the following equation using the *electron density n* and *hole density p*:

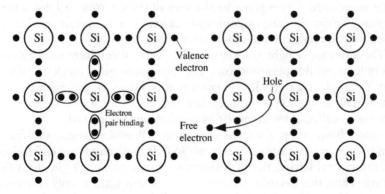

Figure 4.5 *Crystal Structure of Silicon (left), Intrinsic Conduction due to Defect Electron in the Crystal Lattice (right)*

$$n = p \tag{4.13}$$

The product of electron and hole density is called the *intrinsic carrier density* n_i^2 and depends on the absolute temperature T and band gap E_g:

$$n \cdot p = n_i^2 = n_{i0}^2 \cdot T^3 \cdot \exp(-\frac{E_g}{k \cdot T}) \tag{4.14}$$

where the Boltzmann constant k is given as:

$$k = 1.380658 \cdot 10^{-23} \; J/K \tag{4.15}$$

The value for silicon is $n_{i0} = 4.62 \cdot 10^{15} \; cm^{-3} \; K^{-3/2}$. No free electrons and holes exist at a temperature of absolute zero ($T = 0$ K $= -273.15°C$). With increasing temperature their number rises rapidly.

If an electrical voltage is applied to the silicon crystal externally, negatively charged electrons will flow to the anode. Electrons neighbouring a hole can move into the hole created by this current. Thus, holes move in the opposite direction to the electrons. The *mobility* μ_n and μ_p of electrons and holes in the semiconductor depends also on the temperature. μ_n and μ_p can be calculated for silicon with $\mu_{0n} = 1350 \; cm^2/(V \; s)$ and $\mu_{0p} = 480 \; cm^2/(V \; s)$ at $T_0 = 300$ K by:

$$\mu_n = \mu_{0n} \cdot \left(\frac{T}{T_0}\right)^{-3/2} \tag{4.16}$$

$$\mu_p = \mu_{0p} \cdot \left(\frac{T}{T_0}\right)^{-3/2} \tag{4.17}$$

The *electrical conductivity*

$$\kappa = \frac{1}{\rho} = e \cdot (n \cdot \mu_n + p \cdot \mu_p) = e \cdot n_i \cdot (\mu_n + \mu_p) \tag{4.18}$$

of the semiconductors is given by the sum of the electron and hole currents. The conductivity decreases significantly at very low temperatures. This effect is used for the production of low temperature sensors.

The influence of light also changes the electrical conductivity. This effect is used in light sensible photoresistors. For their application, an electrical voltage must be applied externally. However, this effect is not relevant to the photovoltaic generation of electrical current. Therefore, another effect must be used: the so-called extrinsic or *defect conduction* (Figure 4.6).

Atoms from group V such as phosphorus (P) and antimony (Sb) have five valence electrons in contrast to silicon's four. If these atoms are embedded into a silicon crystal lattice, the fifth electron cannot participate in electron pair binding. Thus, this electron is bonded very loosely. Little energy is required to separate this electron from the atom and thus create a free electron. The embedding of atoms from group V is called n-doping. The impurity atoms are called *donors*.

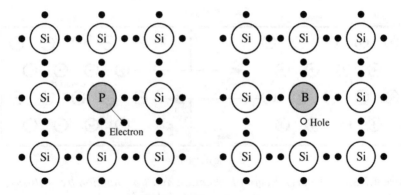

Figure 4.6 *Defect Conduction for n-type and p-type Doped Silicon*

The *density of free electrons* for n-doping is calculated by:

$$n = \sqrt{\frac{n_D \cdot N_L}{2}} \cdot \exp\left(-\frac{E_D}{2 \cdot k \cdot T}\right) \tag{4.19}$$

using the density n_D of donor atoms, the effective state density N_L in the conduction band and the ionization energy E_D necessary to release electrons from the donor atoms. For silicon crystals with a temperature of $T = 300$ K, these values can be estimated as $N_L = 3.22 \cdot 10^{19}$ cm^{-3} and $E_D = 0.044$ eV for phosphorus atoms as donors.

Since n-doped semiconductors have significantly more free electrons than holes, the electrons are called majority carriers. Electrical transport in such materials is nearly exclusively due to an electron current.

Embedding atoms of group III such as boron (B) or aluminium (Al), with three valence electrons, into the silicon lattice will create a missing valance electron and thus holes emerges as the majority carriers. This process is called p-doping; it makes the semiconductor p-type and the impurity atoms are called *acceptors*. A small amount of energy E_A can release a freely moving hole.

With the density of the acceptors n_A, the effective state density N_V in the valence band and the ionizing energy of acceptors E_A ($N_V = 1.83 \cdot 10^{19}$ cm^{-3} for silicon at $T = 300$ K, and $E_A = 0.045$ eV for boron) the *density of free holes* for p-type semiconductors becomes:

$$p = \sqrt{\frac{n_A \cdot N_V}{2}} \cdot \exp\left(-\frac{E_A}{2 \cdot k \cdot T}\right) \tag{4.20}$$

If p-type and n-type semiconductors are placed in contact, a so-called *p-n junction* is created. Owing to the different majority carriers, electrons will diffuse from the n-region into the p-region, and holes from the p-region into the n-region (Figure 4.7).

A region with very few free charge carriers emerges in the border area. Where electrons have diffused into the p-region, positively ionized atoms

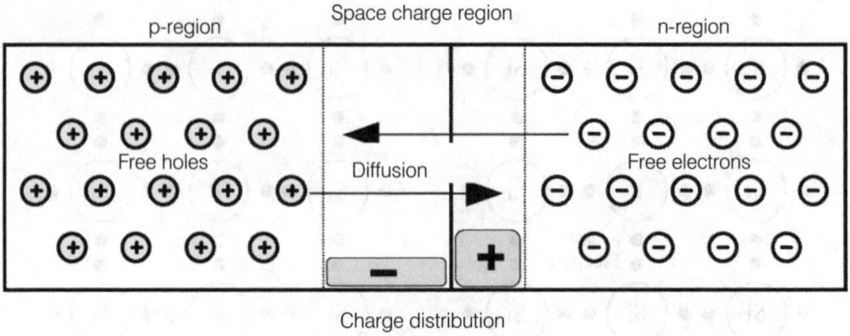

Figure 4.7 *Space Charge Region Formation at a p-n Junction by Diffusion of Electrons and Holes*

remain. They create a positive space charge region. Where holes have diffused into the n-region, negatively ionized atoms remain and create a negative space charge region.

An electric field between the n-region and p-region is thus created. It counteracts the charge carriers and hence the diffusion cannot continue indefinitely. Finally, a *diffusion voltage*

$$V_{d} = \frac{k \cdot T}{e} \cdot \ln \frac{n_{A} \cdot n_{D}}{n_{i}^{2}} \tag{4.21}$$

is created. The charge neutrality within the boundaries d_{n} and d_{p} of the space charge region in the n-type and p-type semiconductor region leads to:

$$d_{n} \cdot n_{D} = d_{p} \cdot n_{A} . \tag{4.22}$$

The total width of the space charge region can then be calculated from:

$$d = d_{n} + d_{p} = \sqrt{\frac{2 \cdot \varepsilon_{r} \cdot \varepsilon_{0} \cdot U_{d}}{e} \cdot \frac{n_{A} + n_{D}}{n_{A} \cdot n_{D}}} . \tag{4.23}$$

For silicon with a dopant concentration of n_{D} = $2 \cdot 10^{16}$ cm^{-3} and n_{A} = $1 \cdot 10^{16}$ cm^{-3} at a temperature of T = 300 K, the diffusion voltage becomes to V_{d} = 0.73 V; and with ε_{r} = 11.8, the widths are d_{n} = 0.13 μm and d_{p} = 0.25 μm.

When electrons are lifted from the valence band into the conduction band and thus released from the atom in the space charge region, the electric field will pull them into the n-region. Similarly, generated holes will move into the p-region. This can be explained in the energy band model by band bending in the space charge region (Figure 4.8).

As described before, the solar cell can only convert a part of the photon energy into electrical current. For photon energies smaller than the band gap, the energy is not sufficient to promote an electron from the valence band to the conduction band. This is the case for wavelengths above:

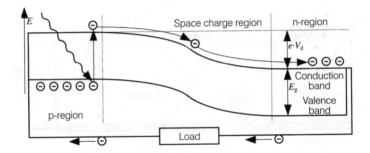

Figure 4.8 *Solar Cell Principle with Energy Band Model*

$$\lambda_{\text{max}} = \frac{h \cdot c}{E_{\text{g}}} = \frac{1.24\,\mu\text{m} \cdot \text{eV}}{E_{\text{g}}} \tag{4.24}$$

Not all the energy of photons with wavelengths near the band gap is converted to electricity. The solar cell surface reflects a part of the incoming light, and some is transmitted through the solar cell. Furthermore, electrons can recombine with holes. In other words, they can fall back to the valence band before they are converted to electricity. Figure 4.9 describes these processes.

The solar cell only uses an amount of energy equal to the band gap of the higher energy of photons with lower wavelengths. Excess energy, i.e. energy above the band gap equivalent, is passed on to the crystal in the form of heat. Hence, the share of the usable energy mainly depends on the wavelength and the band gap. The external quantum collecting efficiency $\eta_{\text{ext}}(\lambda)$ is the likelihood that an incident photon generates an electron–hole pair. It is closely related to the spectral response, which is a measure of the part of the energy converted into charge carriers. In photovoltaics, the *spectral response* $S(\lambda)$ given by:

$$S(\lambda) = \frac{e \cdot \lambda}{h \cdot c} \cdot \eta_{\text{ext}}(\lambda) = \frac{\lambda}{1.24\mu\text{m}} \cdot \frac{\text{A}}{\text{W}} \cdot \eta_{\text{ext}}(\lambda) \tag{4.25}$$

is defined using the external quantum collecting efficiency η_{ext} and the wavelength λ. Figure 4.10 shows the spectral response S as a function of the wavelength λ.

In the absence of an external field, i.e. if a solar cell is short-circuited, the *photocurrent* I_{Ph} is generated. This current can be calculated using the solar cell area A, the spectral sensitivity S and the spectrum of sunlight $E(\lambda)$ (e.g. the air mass AM 1.5 spectrum of Chapter 2):

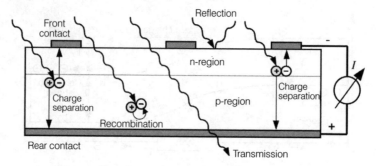

Figure 4.9 *Processes in an Irradiated Solar Cell*

$$I_{Ph} = \int S(\lambda) \cdot E(\lambda) \cdot A \cdot d\lambda \qquad (4.26)$$

The irradiance E absorbed by the semiconductor is a share of the incoming irradiance E_0. It depends on the thickness d of the semiconductor and the material-dependent *absorption coefficient* α:

$$E = E_0 \cdot (1 - \exp(-\alpha \cdot d)) \qquad (4.27)$$

There are two types of semiconductor: direct and indirect. The absorption coefficient of indirect semiconductors such as silicon is significantly lower for higher wavelengths. For example, the direct semiconductor GaAs has an absorption coefficient for light with a wavelength of about 1 μm of $\alpha(\text{GaAs})$ $\approx$ 630 mm^{-1}, whereas this value decreases to $\alpha(\text{Si}) \approx 7.2$ mm^{-1} for silicon. For both semiconductors to absorb the same amount of light, the silicon will have to be 87.5 times thicker than a GaAs semiconductor. The wavelength dependence of the absorption coefficients must be considered for an exact calculation. Crystalline silicon solar cells should have a thickness of at least

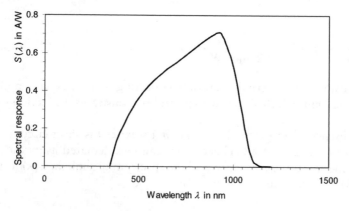

Source: Wagemann and Eschrich, 1994

Figure 4.10 *Spectral Response of a Solar Cell*

about 200 μm for high absorptions. So-called light trapping, which reflects the light in the material, enlarges the path length and reduces the required thickness.

Further physical details of solar cells as well as descriptions of other solar cell technologies such as metal–insulator–semiconductor (MIS) cells are not given here. Details can be found in the literature, for example, Goetzberger et al (1998); Green (1994); Luque and Heqedus (2003); Marti and Luque (2003).

PRODUCTION OF SOLAR CELLS AND SOLAR MODULES

Crystalline silicon solar cells

Various semiconductor materials are suited to solar cell production; however, silicon is the most commonly used material today. For this reason, only the process of producing solar cells from silicon is described here.

Silicon can mainly be found in quartz sand (SiO_2). The following reduction process extracts silicon from the quartz sand at high temperatures of about 1800°C (3272°F):

$$SiO_2 + 2C \xrightarrow{\quad 1800°C \quad} Si + 2CO \tag{4.28}$$

The result of this reaction is so-called *metallurgical-grade silicon* (MG-Si) with a purity of about 98 per cent. Another process for extracting silicon is the aluminothermic reduction:

$$3SiO_2 + 4Al \xrightarrow{\quad 1100°C–1200°C \quad} 2Al_2O_3 + 3Si \tag{4.29}$$

However, silicon gained by this process also has significant impurities. Silicon used by the computer industry is so-called *electronic-grade silicon* (EG-Si) for the production of semiconductor devices. Its impurity level is below 10^{-10} per cent. This high purity is not necessary for solar cell production, in which *solar-grade silicon* (SOG-Si) is commonly used. Nevertheless, purification processes are needed for the production of SOG-Si.

Silicon is mixed with hydrogen chloride or chloric acid (HCl) in the *silane process*. An exothermic reaction produces trichlorosilane ($SiHCl_3$) and hydrogen (H_2):

$$Si + 3HCl \longrightarrow SiHCl_3 + H_2 \tag{4.30}$$

Trichlorosilane is liquid at temperatures of 30°C. Multiple fraction distillations are used to remove the impurities. The chemical vapour deposition (CVD) process is used for silicon recovery. Silicon is deposited as a thin silicon rod at temperatures of 1350°C (2462°F), when the trichlorosilane is brought into contact with high-purity hydrogen:

$$4\,SiHCl_3 + 2\,H_2 \xrightarrow{\quad 1350°C \quad} 3\,Si + SiCl_4 + 8\,HCl \qquad (4.31)$$

The end product is a high-purity silicon rod with diameters of up to 30 cm (about 12 inches) and lengths up to 2 m (about 80 inches). These rods can be used for the production of polycrystalline solar cells, which consist of a number of crystals, rather than a single crystal. The crystals of polycrystalline silicon are differently oriented and separated by grain boundaries. They introduce some efficiency losses.

To increase solar cell efficiency, monocrystalline material can be produced from polycrystalline material applying the Czochralski or float zone process. Seeding a single crystal at high temperatures transforms the polycrystalline silicon to the desired monocrystalline silicon. No grain boundaries are present in the resulting material and thus losses within a solar cell are reduced.

Wire saws or inner diameter saws cut the crystalline silicon rods into 200-μm to 500-μm slices. This process causes relatively high cutting losses of up to 50 per cent.

The silicon slices, or so-called *wafers*, are cleaned and doped in the following steps. Hydrofluoric acid removes any saw damage. Phosphorus and boron are used for doping silicon to create the p-n junction. Gaseous dopants are mixed with a carrier gas such as nitrogen (N_2) or oxygen (O_2) for gas diffusion, and this gas mixture flows over the silicon wafers. The impurity atoms diffuse into the silicon wafer depending on the gas mixture, temperature and flow velocity. Etching cleans the surface of the doped semiconductor.

Finally, cell contacts are added. A screen printing process adds the *front and rear contacts*. Materials for the contacts are metals or alloys of aluminium or silver. The rear contact usually covers the whole cell area. Thin contact fingers are used for the front contacts, because they obstruct and reflect sunlight. Only a minimum of the cell's surface should be covered by contacts in order to optimize light capture.

Finally, an *antireflective coating* is added to the solar cell. This coating reduces reflection at the metallic silicon surface. Titanium dioxide (TiO_2) is mostly used for the coating and gives the solar cell its typical blue colour. Nowadays, it is also possible to produce antireflective coatings in other colours, allowing architects to better integrate solar modules with buildings. Figure 4.11 shows the structure of a crystalline solar cell.

Various other methods can be employed to increase the efficiency. For example, the solar cell's surface can be structured with miniature pyramids. The pyramids are shaped in such a way that any reflection of the light is directed onto the cell, hence producing a second incident beam. Furthermore, buried front contacts can reduce the reflection losses. A more detailed description of the production methods can be found in Goetzberger et al, 1998; Green, 1994; Lasnier and Ang, 1990.

Solar modules with crystalline cells

Single unprotected solar cells can be damaged rapidly as a result of climatic

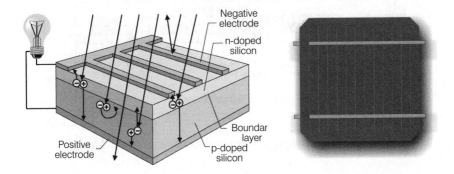

Figure 4.11 *Solar Cell Structure and Front View of a Crystalline Silicon Solar Cell*

influences. To avoid this, further protection is necessary. Several solar cells with an edge length between 10 and 21 cm (4–8 inches) are combined in a solar *module* for cell protection. Many modules are made up of 32–40 cells; however, other module sizes with significantly more or fewer cells exist. The front cover is formed by low-iron glass, which has already been described for the flat solar thermal collector in Chapter 3. The back cover is made of glass or plastic. Between the front and rear cover, the solar cells are embedded into plastic. This material is usually ethylene vinyl acetate (EVA), which is cured at temperatures of 100°C (212°F) at reduced pressure. This process is called *lamination*. A frame is added to the finished modules in some cases. A junction box protects the contacts from water; bypass diodes are also mounted in this box (see section headed 'Series connection under inhomogeneous conditions', p143).

Technical data for solar modules are given in Table 4.6.

Thin film modules

Besides crystalline silicon, thin film modules hold promise for the cells of the future. They can be made of amorphous silicon and other materials such as cadmium telluride (CdTe) or copper indium diselenide ($CuInSe_2$ or CIS). Thin film modules can be produced using a fraction of the semiconductor material necessary for crystalline modules and this promises lower production costs in the medium term. Therefore, the development potential of thin film modules is very high. A disadvantage is the use of toxic materials in some cells; therefore, the importance of production safety and material recycling will increase in the coming years. Furthermore, the supplies of certain materials, for example of tellurium, are restricted, whereas silicon reserves are almost inexhaustible. However, it is not yet clear which material will dominate future markets. This chapter will give only a short illustrative description of amorphous silicon thin film modules because they are the thin film modules with the longest market history.

The base for *amorphous silicon solar modules* is a substrate; in most cases

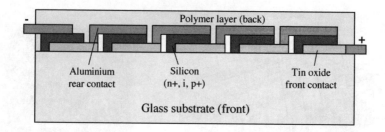

Figure 4.12 *Structure of an Amorphous Silicon Solar Module*

this is glass or a metal foil. A spray process deposits a thin layer of transparent tin oxide on the glass substrate. This layer serves as the transparent front contact. A laser cuts this layer into strips in order to create an integrated series connection and vapour deposition at high temperatures adds silicon and dopants to the substrate. A 10-nm p-layer and afterwards a 10-nm buffer layer are deposited. An intrinsic 1000-nm layer of amorphous silicon and finally a 20-nm n-layer follow. Screen-printing processes add the back contacts, which are mostly made of aluminium. The samples are then either laminated or coated with a polymer to protect the solar module from climatic influences. Figure 4.12 shows the principle structure of an amorphous silicon solar module.

Silicon loses its crystalline structure during vapour depositing in amorphous solar cells. The main advantage of amorphous silicon cells is that they are thinner than crystalline cells by a factor of 100. This is only possible because amorphous silicon is a direct semiconductor with a much higher absorption coefficient than crystalline silicon. Due to its amorphous structure, the band gap is higher than crystalline silicon, at 1.7 eV. The production process saves a lot of material and even more efficient production is possible. Laboratory efficiencies of 13 per cent have been reached. However, amorphous silicon solar cells are mainly used in small applications such as pocket calculators or wristwatches largely because the efficiency achieved in production is only around 6–8 per cent, which is much lower than that of crystalline cells, which reach up to 20 per cent. Furthermore, amorphous silicon solar cells exhibit a degradation process. This reduces the efficiency in the first months of operation by 10–20 per cent until the performance finally stabilizes.

Other thin film materials already have started to be commercialized. Various materials and technologies are currently under development. Besides the use of new materials such as cadmium telluride (CdTe) or copper indium diselenide (CuInSe$_2$), microcrystalline thin film cells are under development. Other promising developments are dye-sensitized cells on a base of titanium dioxide (TiO$_2$). Which technology will become commercially viable depends on further technological developments.

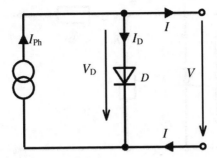

Figure 4.13 *Simple Equivalent Circuit of a Solar Cell*

ELECTRICAL DESCRIPTION OF SOLAR CELLS

Simple equivalent circuit

A photovoltaic solar cell is a large area *diode*. It consists of an n-type and p-type doped semiconductor with a resulting space charge layer. Typically, a non-irradiated solar cell has nearly the same behaviour as a diode. Therefore, a simple diode can describe the equivalent circuit.

The equation of the cell current I depends on the cell voltage (here $V = V_D$) with the saturation current I_S and the diode factor m:

$$I = -I_D = -I_S \cdot \left(\exp\left(\frac{V_D}{m \cdot V_T} \right) - 1 \right) \tag{4.32}$$

The thermal voltage V_T at a temperature of 25°C is $V_T = 25.7$ mV. The magnitude of the saturation current I_S is of the order of 10^{-10}–10^{-5}A. The

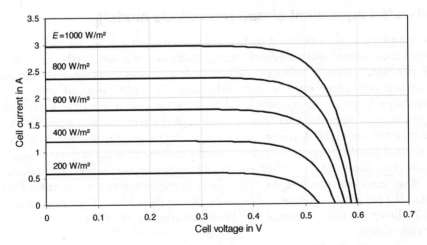

Figure 4.14 *Influence of the Irradiance* E *on the I-V Characteristics of a Solar Cell*

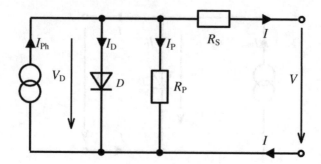

Figure 4.15 *Extended Equivalent Circuit of a Solar Cell (One-diode Model)*

diode factor m of an ideal diode is equal to 1; however, a diode factor between 1 and 5 allows a better description of the solar cell characteristics.

A current source connected in parallel to the diode completes the simple equivalent circuit of an irradiated solar cell. The current source generates the *photocurrent* I_{ph}, which depends on the irradiance E and the coefficient c_0:

$$I_{Ph} = c_0 \cdot E \tag{4.33}$$

Kirchhoff's first law provides the current–voltage characteristics of the simple solar cell equivalent circuit illustrated in Figures 4.13, and Figure 4.14 shows the characteristic curves at different irradiances):

$$I = I_{Ph} - I_D = I_{Ph} - I_S \cdot \left(\exp\left(\frac{V}{m \cdot V_T} \right) - 1 \right) \tag{4.34}$$

Extended equivalent circuit (one-diode model)

The simple equivalent circuit is sufficient for most applications. The differences between calculated and measured characteristics of real solar cells are only a few per cent. However, only extended equivalent circuits describe the electrical solar cell behaviour exactly, especially when a wide range of operating conditions is to be investigated. Charge carriers in a realistic solar cell experience a voltage drop on their way through the semiconductor junction to the external contacts. A *series resistance* R_S expresses this voltage drop. An additional *parallel resistance* R_P describes the leakage currents at the cell edges. Figure 4.15 shows the modified equivalent circuit including both resistances.

The series resistance R_S of real cells is in the range of several milliohms (mΩ), the parallel resistance R_P is usually higher than 10 Ω. Figures 4.16 and 4.17 illustrate the influence of both resistances in terms of the I-V characteristics.

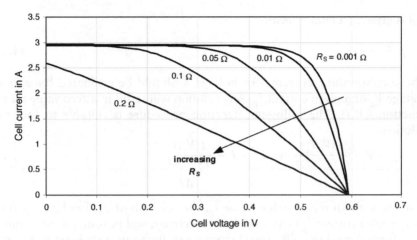

Figure 4.16 *Influence of the Series Resistance* R$_S$ *on the I-V Characteristics of a Solar Cell*

Kirchhoff's nodal law, $0 = I_{Ph} - I_D - I_P - I$ with $I_P = \dfrac{V_D}{R_P} = \dfrac{V + I \cdot R_S}{R_P}$, provides the equation for the I-V characteristics of the extended solar cell equivalent circuit:

$$0 = I_{Ph} - I_S \cdot \left(\exp\left(\frac{V + I \cdot R_S}{m \cdot V_T} \right) - 1 \right) - \frac{V + I \cdot R_S}{R_P} - I .$$

(4.35)

This implicit equation cannot be solved as easily as Equation (4.34) for the current *I* or voltage *V*. Numerical methods are therefore needed.

One common method for solving this equation is by so-called root finding methods, estimating solutions at which the above equation is zero. What will be described here is the Newton method. The I-V characteristic of the solar

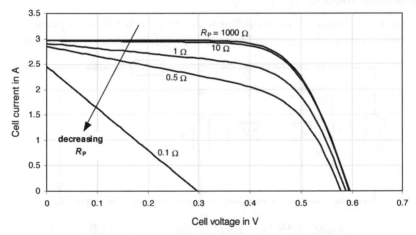

Figure 4.17 *Influence of the Parallel Resistance* R$_P$ *on the I-V Characteristics of a Solar Cell*

cell is given in a closed form:

$$f(V,I) = 0 \tag{4.36}$$

The corresponding current I or voltage V should be estimated for a given voltage V_g or a given current I_g. Any solution will result in a zero value for the function $f(V,I)$. The following iteration procedure is suitable to find this solution:

$$V_{i+1} = V_i - \frac{f(V_i,I_g)}{\dfrac{\mathrm{d}f(V_i,I_g)}{\mathrm{d}V}} \quad \text{or} \quad I_{i+1} = I_i - \frac{f(V_g,I_i)}{\dfrac{\mathrm{d}f(V_g,I_i)}{\mathrm{d}I}} \tag{4.37}$$

Starting with an initial value V_0 or I_0, the solution of the implicit equation with a given current I_g or voltage V_g, respectively, will be found if the iteration is performed until the difference between two iteration steps is smaller than a predefined limit ε. The stopping conditions for the iteration are: $|V_i - V_{i-1}| < \varepsilon$ or $|I_i - I_{i-1}| < \varepsilon$.

The Newton method tends to convergence very quickly; however, the speed of convergence depends strongly on the chosen initial value V_0 or I_0. A pre-iteration using another method could be useful near the range of the diode breakdown voltage.

The iteration for estimating the current I of the solar cell for a given voltage V_g according to Equation 4.37 is:

$$I_{i+1} = I_i - \frac{I_{Ph} - I_S \cdot \left(\exp\left(\dfrac{V_g + I_i \cdot R_S}{m \cdot V_T} \right) - 1 \right) - \dfrac{V_g + I_i \cdot R_S}{R_P} - I_i}{-\dfrac{I_S \cdot R_S}{m \cdot V_T} \cdot \exp\left(\dfrac{V_g + I_i \cdot R_S}{m \cdot V_T} \right) - \dfrac{R_S}{R_P} - 1} \tag{4.38}$$

Two-diode model

The two-diode (Figure 4.18) model provides an even better description of the solar cell in many cases. A second diode is therefore connected in parallel to

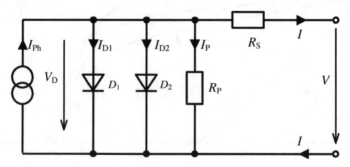

Figure 4.18 *Two-diode Model of a Solar Cell*

Table 4.3 *Two-diode Parameters for Various Photovoltaic Modules*

Parameter	c_0	I_{S1}	I_{S2}	m_1	m_2	R_S	R_P
Unit	m^2/V	nA	μA	–	–	$m\Omega$	Ω
AEG PQ 40/50	$2.92 \cdot 10^{-3}$	1.082	12.24	1	2	13.66	34.9
Siemens M50	$3.11 \cdot 10^{-3}$	0.878	12.71	1	2	13.81	13.0
Kyocera LA441J59	$3.09 \cdot 10^{-3}$	1.913	8.25	1	2	12.94	94.1

Source: University of Oldenburg, 1994

the first diode. Both diodes have different saturation currents and diode factors. Thus, the implicit equation for the two-diode model becomes:

$$0 = I_{Ph} - I_{S1} \cdot \left(\exp\left(\frac{V + I \cdot R_S}{m_1 \cdot V_T} \right) - 1 \right) - I_{S2} \cdot \left(\exp\left(\frac{V + I \cdot R_S}{m_2 \cdot V_T} \right) - 1 \right) - \frac{V + I \cdot R_S}{R_P} - I$$

$$(4.39)$$

The first diode is usually an ideal diode ($m_1 = 1$). The diode factor of the second diode is $m_2 = 2$. Table 4.3 summarizes parameters that have demonstrated good simulation results for some modules.

Two-diode model with extension term

The equivalent circuit of the solar cell must be extended for the description of the negative breakdown characteristics at high negative voltages to be able to model the complete I-V characteristics. The additional current source $I(V_D)$ in Figure 4.19 expresses the extension term, which describes the diode breakdown in the negative voltage range.

This current source generates a current depending on the diode voltage V_D. This current describes the electrical solar cell behaviour at negative voltages. With the breakdown voltage V_{Br}, the avalanche breakdown exponent n and the correction conductance b, the equation for the I-V characteristics becomes:

$$0 = I_{Ph} - I_{S1}\left(\exp\left(\frac{V + I \cdot R_S}{m_1 \cdot V_T} \right) - 1 \right) - I_{S2}\left(\exp\left(\frac{V + I \cdot R_S}{m_2 \cdot V_T} \right) - 1 \right) - \frac{V + I \cdot R_S}{R_P} -$$

$$\underbrace{- b \cdot (V + I \cdot R_S) \cdot \left(1 - \frac{V + I \cdot R_S}{V_{Br}} \right)^{-n}}_{\text{expansion term}}$$

$$(4.40)$$

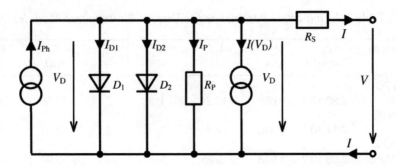

Figure 4.19 *Two-diode Equivalent Circuit with Second Current Source to Describe the Solar Cell Breakdown at Negative Voltages*

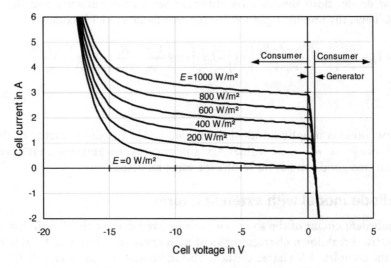

Figure 4.20 *I-V Characteristics of a Polycrystalline Solar Cell over the Full Voltage Range*

Figure 4.20 shows the I-V characteristics of a polycrystalline solar cell over the full voltage range obtained with the parameters $I_{S1} = 3 \cdot 10^{-10}$ A, $m_1 = 1$, $I_{S2} = 6 \cdot 10^{-6}$ A, $m_2 = 2$, $R_S = 0.13$ Ω, $R_P = 30$ Ω, $V_{Br} = -18$ V, $b = 2.33$ mS and $n = 1.9$. In this instance, the series resistance is relatively high due to the inclusion of the connections. In the given figure, cell voltage and current are positive in the quadrant where the solar cell is generating power. If the cell voltage or current becomes negative, the solar cell is operated as load. Therefore, an external voltage source or other solar cells must generate the electrical power required.

Further electrical solar cell parameters

Besides the described correlations of solar cell current and voltage, further characteristic parameters can be defined. This section describes the most common parameters.

The voltage of a short-circuited solar cell is equal to zero, in which case, the short circuit current I_{SC} is approximately equal to the photocurrent I_{Ph}. Since the photocurrent is proportional to the irradiance E, the *short circuit current* also depends on the irradiance:

$$I_{SC} \approx I_{Ph} = c_0 \cdot E$$

The short circuit current rises with increasing temperature. The standard temperature for reporting short circuit currents I_{SC} is usually $\vartheta = 25°C$. The temperature coefficient α_{ISC} of the short circuit current allows its value to be estimated at other temperatures:

$$I_{SC}(\vartheta_2) = I_{SC}(\vartheta_1) \cdot (1 + \alpha_{ISC} \cdot (\vartheta_2 - \vartheta_1)) \tag{4.41}$$

For silicon solar cells, the temperature coefficient of the short circuit current is normally between $\alpha_{ISC} = +10^{-3}/°C$ and $\alpha_{ISC} = +10^{-4}/°C$.

If the cell current I is equal to zero, the solar cell is in open circuit operation. The cell voltage becomes the *open circuit voltage* V_{OC}. The I-V equation of the simple equivalent circuit, see Equation (4.34), provides V_{OC} when setting I to zero:

$$V_{OC} = m \cdot V_T \cdot \ln\left(\frac{I_{SC}}{I_s} + 1\right) \tag{4.42}$$

Since the short circuit current I_{SC} is proportional to the irradiance E, the open circuit voltage dependence is:

$$V_{OC} \sim \ln(E) \tag{4.43}$$

The temperature coefficient α_{VOC} of the open circuit voltage is obtained similarly to the short circuit current. It commonly has a negative sign. For silicon solar cells, the temperature coefficient is between $\alpha_{VOC} = -3 \cdot 10^{-3}/°C$ and $\alpha_{VOC} = -5 \cdot 10^{-3}/°C$. In other words, the open circuit voltage decreases faster with rising temperature than the short circuit current increases (see section on temperature dependence, p139).

The solar cell generates maximum power at a certain voltage. Figure 4.21 shows the current–voltage as well as the power–voltage characteristic. It shows clearly that the power curve has a point of maximal power. This point is called the *maximum power point (MPP)*.

The voltage at the MPP, V_{MPP}, is less than the open circuit voltage V_{OC}, and the current I_{MPP} is lower than the short circuit current I_{SC}. The MPP current and voltage have the same relation to irradiance and temperature as the short circuit current and open circuit voltage. The maximum power P_{MPP} is:

$$P_{MPP} = V_{MPP} \cdot I_{MPP} < V_{OC} \cdot I_{SC} \tag{4.44}$$

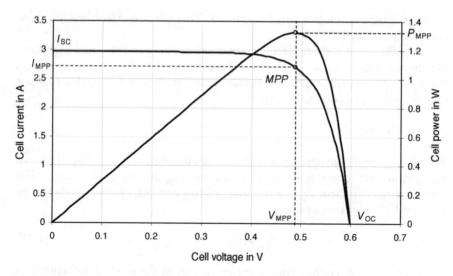

Figure 4.21 *I-V and P-V Solar Cell Characteristics with Maximum Power Point (MPP)*

Since the temperature coefficient of the voltage is higher than that of the current, the temperature coefficient α_{PMPP} of the MPP power is negative. For silicon solar cells it is between $\alpha_{PMPP} = -3 \cdot 10^{-3}/°C$ and $\alpha_{PMPP} = -6 \cdot 10^{-3}/°C$. A temperature rise of 25°C causes a power drop of about 10 per cent.

In order to make possible comparisons between solar cells and modules , MPP power is measured under *standard test conditions* (STC) (E = 1000 W/m², ϑ = 25°C, AM 1.5). The generated power of the solar modules under real weather conditions is usually lower. Hence STC power has the unit W_p (Watt-peak).

Considering the irradiance dependence, the current dominates the device behaviour, so that the MPP power is nearly proportional to the irradiance E.

Another parameter is the *fill factor* (FF) with the definition:

$$FF = \frac{P_{MPP}}{V_{OC} \cdot I_{SC}} = \frac{V_{MPP} \cdot I_{MPP}}{V_{OC} \cdot I_{SC}} \tag{4.45}$$

The fill factor is a quality criterion of solar cells that describes how well the I-V curve fits into the rectangle of V_{OC} and I_{SC}. The value is always smaller than 1 and is usually between 0.75 and 0.85.

Together, the MPP power P_{MPP}, the irradiance E and the solar cell area A provide the solar cell *efficiency* η:

$$\eta = \frac{P_{MPP}}{E \cdot A} = \frac{FF \cdot V_{OC} \cdot I_{SC}}{E \cdot A} \tag{4.46}$$

The efficiency is usually determined under standard test conditions. Table 4.4 summarizes the various solar cell parameters.

Table 4.4 *Electrical Solar Cell Parameters*

Name	Symbol	Unit	Remarks
Open circuit voltage	V_{OC}	V	$V_{OC} \sim \ln E$
Short circuit current	I_{SC}	A	$I_{SC} \approx I_{Ph} \sim E$
MPP voltage	V_{MPP}	V	$V_{MPP} < V_{OC}$
MPP current	I_{MPP}	A	$I_{MPP} < I_{SC}$
MPP power	P_{MPP}	W or W_p	$P_{MPP} = V_{MPP} \cdot I_{MPP}$
Fill factor	FF		$FF = P_{MPP} / (V_{OC} \cdot I_{SC}) < 1$
Efficiency	η	%	$\eta = P_{MPP} / (E \cdot A)$

Temperature dependence

A constant temperature of 25°C was assumed for all equations of the previous section. It was mentioned that the characteristics change with the temperature. This section describes how to modify the solar cell equations to include the temperature dependence.

The thermal voltage V_T must be calculated for a given temperature. With the Boltzmann constant $k = 1.380658 \cdot 10^{-23}$ J/K, the absolute temperature T in Kelvin ($T = \vartheta$ K/°C + 273.15 K) and the elementary charge $e = 1.60217733 \cdot 10^{-19}$ A s, the temperature voltage is given by:

$$V_T = \frac{k \cdot T}{e} . \tag{4.47}$$

The temperature dependence of the saturation currents I_{S1} and I_{S2} with the coefficients c_{S1} and c_{S2} and the band gap E_g of Table 4.2 is given by the following equations (Wolf et al, 1977) for silicon devices:

$$I_{S1} = c_{S1} \cdot T^3 \cdot \exp\left(-\frac{E_g}{k \cdot T}\right) \tag{4.48}$$

$$I_{S2} = c_{S2} \cdot T^{5/2} \cdot \exp\left(-\frac{E_g}{2 \cdot k \cdot T}\right) \tag{4.49}$$

The increase of the saturation currents with rising temperature explains the decrease in the open circuit voltage. The temperature dependence of the series resistance R_S, the parallel resistance R_p and the diode factor can be ignored for further considerations.

Equations 4.48 and 4.49 ignore the temperature dependence of the bad gap. While it does not significantly influence the saturation currents, its temperature dependence is decisive for the *photocurrent* I_{Ph}. Due to the decrease in the bad gap with rising temperature, photons with lower energy can elevate electrons into the valence band, which increases the photocurrent. Using the coefficients c_1 and c_2, the temperature dependence of the photocurrent is given by:

Table 4.5 *Parameters for the Temperature Dependence of Various Photovoltaic Modules*

	Parameter			
Unit	c_{S1} $A \ / \ K^3$	c_{S2} $A \ K^{-5/2}$	c_1 m^2/V	c_2 $m^2/(V \ K)$
AEG PQ 40/50	210.4	$18.1 \cdot 10^{-3}$	$2.24 \cdot 10^{-3}$	$2.286 \cdot 10^{-6}$
Siemens M50	170.8	$18.8 \cdot 10^{-3}$	$3.06 \cdot 10^{-3}$	$0.179 \cdot 10^{-6}$
Kyocera LA441J59	371.9	$12.2 \cdot 10^{-3}$	$2.51 \cdot 10^{-3}$	$1.932 \cdot 10^{-6}$

Source: University of Oldenburg, 1994

$$I_{Ph}(T) = (c_1 + c_2 \cdot T) \cdot E \tag{4.50}$$

Table 4.5 shows the parameters for the calculation of the temperature dependences of various solar modules.

Figure 4.22 shows the I-V characteristic with rising temperature ϑ. It shows clearly that the open circuit voltage decreases significantly when the temperature rises. On the other hand, the short circuit current only increases slightly. The result is the reduction of the MPP power at decreasing temperatures.

Parameter estimation

For the simulation of a real solar cell, for instance with the simple equivalent circuit, the cell parameters (here I_{Ph} and I_S) must be estimated from measured cell characteristics. To simplify the process, the photocurrent I_{Ph} can be assumed to be equal to the short circuit current I_{SC}. The diode factor m of an ideal diode is equal to 1. Hence two parameters are already estimated ($I_{Ph} = I_{SC}$ and $m = 1$). The *diode saturation current* I_S can be calculated from equation (4.34) using the open circuit conditions:

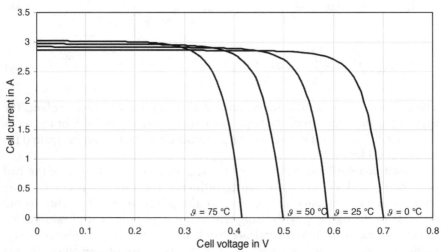

Figure 4.22 *Temperature Dependence of Solar Cell Characteristics*

$$I_S = \frac{I_{SC}}{\exp\left(\dfrac{V_{OC}}{V_T}\right) - 1} \approx I_{SC} \cdot \exp\left(-\frac{V_{OC}}{V_T}\right) \tag{4.51}$$

With this, all parameters of the simple equivalent circuit are estimated. However, this model can only provide a very rough correspondence with measured characteristics. A diode factor higher than 1 ($m > 1$) is used for a non-ideal diode. The two independent parameters m and I_S could be found relatively easy using mathematical software such as Mathematica for a given solar cell characteristic in the generating region. The simple equivalent circuit with a real diode already provides a very good correspondence between simulations and measurements.

The determination of the additional parameters R_S and R_P of the extended one-diode solar cell model is more difficult. With higher numbers of independent parameters, even professional mathematical programs reach their limits and the best fit has to be determined iteratively. The initial values must be chosen carefully for a good iteration convergence of the parameter estimation. However, the estimation of initial values for R_P and R_S is relatively simple.

The *parallel resistance* R_P can be estimated by the negative slope of the I-V characteristic under short circuit conditions. The slope of the I-V characteristic around or beyond the open circuit voltage provides the *series resistance* R_S:

$$R_P \approx \left.\frac{\partial V}{\partial I}\right|_{V=0} \tag{4.52}$$

$$R_S \approx \left.\frac{\partial V}{\partial I}\right|_{V \gg V_{OC}} \tag{4.53}$$

The parameters V_{Br}, b and n for the negative diode breakdown operation can be found analogously to the other parameters, when measured values at the point of negative diode breakdown are used for parameter estimation.

ELECTRICAL DESCRIPTION OF PHOTOVOLTAIC MODULES

Series connection of solar cells

Solar cells are normally not operated individually due to their low voltage. In photovoltaic modules, cells are mostly connected in series. A connection of these modules in series, parallel or series–parallel combinations builds up the photovoltaic system.

Many modules are designed for operation with 12-V lead–acid rechargeable batteries where a series connection of 32–40 silicon cells is optimal. Modules for grid connection can have many more cells connected in series in order to obtain higher voltages.

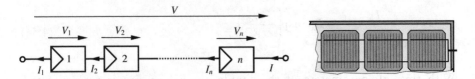

Figure 4.23 *Series Connection of Photovoltaic Solar Cells (left: Electrical Symbols, Currents and Voltages; right: Top View of a Part of a Module with Crystalline Cells)*

The current I_i through all cells i of a series connection of n cells is identical, according to Kirchhof's law (see Figure 4.23). The cell voltages V_i are added to obtain the overall module voltage V:

$$I = I_1 = I_2 = \dots = I_n \tag{4.54}$$

$$V = \sum_{i=1}^{n} V_i \tag{4.55}$$

Given that all cells are identical and experience the same irradiance and temperature, the total voltage is given as:

$$V = n \cdot V_i \tag{4.56}$$

The characteristics of a single cell provide easily the I-V characteristic for any series connection as shown in Figure 4.24.

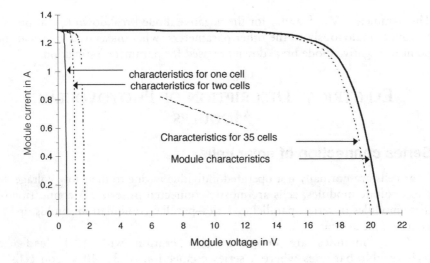

Figure 4.24 *Construction of Module Characteristics with 36 Cells (Irradiance E = 400 W/m², T = 300 K)*

Data sheets published by the module producers often give only a limited number of parameters such as open circuit voltage V_{OC0}, short circuit current I_{SC0}, voltage V_{MPP0} and current I_{MPP0} at the MPP at an irradiance of $E_{1000} = 1000$ W/m^2 and a temperature of $\vartheta_{25} = 25°C$ as well as the temperature coefficients α_V and α_I for voltage and current. The equations:

$$V_{OC} = V_{OC0} \cdot \ln(E) / \ln(E_{1000}) \cdot (1 + \alpha_V(\vartheta - \vartheta_{25})) \qquad (4.57)$$

$$V_{MPP} = V_{MPP0} \cdot \ln(E) / \ln(E_{1000}) \cdot (1 + \alpha_V(\vartheta - \vartheta_{25})) \qquad (4.58)$$

$$I_{MPP} = I_{MPP0} \cdot E / E_{1000} \cdot (1 + \alpha_I(\vartheta - \vartheta_{25})) \qquad (4.59)$$

$$I_{SC} = I_{SC0} \cdot E / E_{1000} \cdot (1 + \alpha_I(\vartheta - \vartheta_{25})) \qquad (4.60)$$

allow fast approximate module performance parameter estimation for different temperatures ϑ and irradiances E. With the parameters

$$c_1 = I_{SC} \cdot \exp(-c_2 \cdot V_{OC}) \qquad (4.61)$$

$$c_2 = \ln(1 - I_{MPP} / I_{SC}) / (V_{MPP} - V_{OC}) \qquad (4.62)$$

the relation between module current I and module voltage V can be found approximately as:

$$I = I_{SC} - c_1 \cdot \exp(c_2 \cdot V) \qquad (4.63)$$

Series connection under inhomogeneous conditions

During realistic operation, not all solar cells of a series connection experience the same conditions. Soiling by leaves or bird excrement, or climatic influences such as snow covering or visual obstructions by surroundings can shadow some cells. This has a high influence on the module characteristics.

Modelling modules with non-identical I-V cell characteristics is significantly more difficult. The following example assumes that 35 cells of a module with 36 series-connected cells are irradiated identically (Quaschning and Hanitsch, 1996). The remaining cell experiences an irradiance reduced by 75 per cent. Even in this case the current through all cells is the same. The module characteristics can be found by choosing a range of currents from zero to the unshaded short circuit current. The voltages of the fully irradiated cells V_F and the shaded cell V_S are determined and added:

$$V = V_S(I) + 35 \cdot V_F(I) \qquad (4.64)$$

The module characteristic can be obtained when stopping at the short circuit current of the partially shaded cell. However, this characteristic only covers a small voltage range close to the module's open circuit voltage. Lower voltage

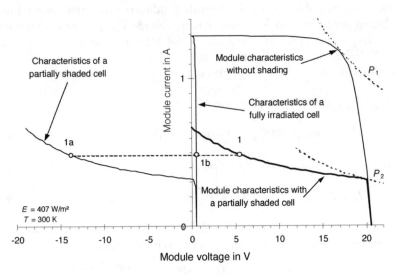

Figure 4.25 *Construction of Module Characteristics with a 75 per cent Shaded Cell*

ranges of this characteristic only occur if the current in the partially shaded cell is higher than the cell short circuit current. This is only possible in the negative voltage range of the shaded cell, and this cell then operates as a load that can be described by the equivalent circuit shown in Figure 4.19.

Figure 4.25 shows the determination of one point of the module characteristic (1). The module voltage for a given current is the sum of the voltage of the partially shaded cell (1a) and 35 times the voltage of the irradiated cells (1b). The total module characteristic of the shaded case is calculated this way point by point for different currents.

It is obvious that *cell shading* reduces the module performance drastically. The maximum module power decreases from P_1 = 20.3 W to P_2 = 6.3 W, i.e. by about 70 per cent, although only 2 per cent of the module surface is shaded. The partially shaded solar cell operates as a load in this example. The dissipated power of the shaded cell is 12.7 W and is obtained when the module is short circuited.

Other shading situations at higher irradiances can increase the power dissipated in the shaded cell up to 30 W. This will heat the cell significantly and may even destroy it. So-called *hot spots*, i.e. hot areas about a millimetre in size, can occur where the cell material melts or the cell encapsulation is damaged.

To protect single cells from hot spot related thermal damage, so-called *bypass diodes* are integrated into the solar modules in parallel to the solar cells. These diodes are not active during regular operation, but when shading occurs, a current flows through the diodes. Hence, the integration of bypass diodes eliminates the possibility of high negative voltages, and in the process eliminates the increase in cell temperature of shaded solar cells.

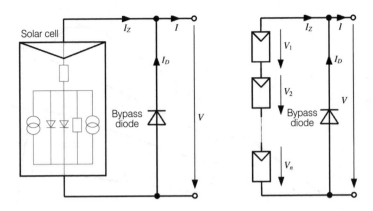

Figure 4.26 *Integration of Bypass Diodes across Single Cells or Cell Strings*

Bypass diodes are usually connected across strings of 18–24 cells. The reason for this is mainly economic. Two bypass diodes are sufficient for a solar module with a rated power of about 50 W containing 36–40 cells. The diodes can be integrated into the module frame or module junction box. However, these diodes cannot fully protect every cell; only the use of one bypass diode for every cell can provide optimal protection. Semiconductor technology can integrate bypass diodes directly into the cells (Hasyim et al, 1986). *Shading-tolerant modules* with cell-integrated bypass diodes, which were first manufactured by Sharp, show significantly lower losses when they are inhomogenously irradiated.

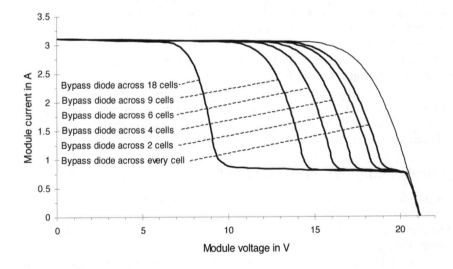

Figure 4.27 *Simulation of Module Characteristics with Bypass Diodes across Different Numbers of Cells (E = 1000 W/m², T = 300 K)*

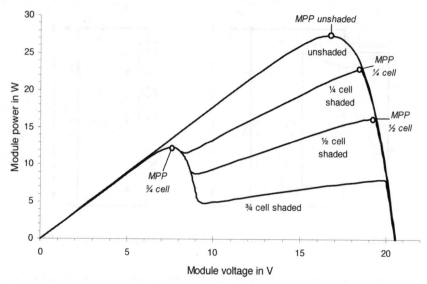

Figure 4.28 *P-V Characteristic of a Module with 36 Cells and Two Bypass Diodes. A Single Cell is Shaded to Different Degrees; All Other Cells Are Fully Irradiated (E = 574 W/m², T = 300 K)*

Figure 4.26 illustrates bypass diode integration across cells and strings of cells. The bypass diode switches as soon as a small negative voltage of about –0.7 V is applied, depending on the type of diode. This negative voltage occurs if the voltage of the shaded cell is equal to the sum of the voltages of the irradiated cells plus that of the bypass diode.

Figure 4.27 shows the shape of I-V characteristics with bypass diodes across a varying number of cells. In this example, one cell is 75 per cent shaded. It is obvious, that the significant drop in the I-V characteristics moves towards higher voltages with decreasing number of cells per bypass diode. This occurs because the bypass diode switches earlier. It also reduces the power loss and the strain on singles cells.

Figure 4.28 shows the power–voltage characteristics of a module with two bypass diodes across 18 cells for different shading situations. Depending on the degree of shading, the MPP shifts and high losses occur although bypass diodes are integrated.

Parallel connection of solar cells

A parallel connection of solar cells is also possible. Parallel connections are less often used than series connections because the associated current increase results in higher transmission losses. Therefore, this section gives only a rough overview on parallel connection.

Parallel-connected solar cells as shown in Figure 4.29 all have the same voltage V. The cell currents I_i are added to obtain the overall current I:

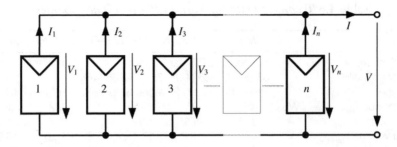

Figure 4.29 *Parallel Connection of* n *Solar Cells*

$$V = V_1 = V_2 = \ldots = V_n \tag{4.65}$$

$$I = \sum_{i=1}^{n} I_i \tag{4.66}$$

A parallel connection of cells is significantly less susceptible to partial shading, and problems associated with cell damage are much less likely.

Large solar generators often use modules with parallel cell strings of multiple series-connected solar cells. However, parallel bypass diodes must also secure cell protection for these modules. So-called serial-connected *blocking diodes* will protect cell strings; however, because of their high permanent diode losses and low protective action, they are used in only very few cases.

Technical data of solar modules

Table 4.6 summarizes technical data for selected solar modules. To point out differences in module technology, some modules are included that are no longer available.

This table contains three modules with monocrystalline cells, one with polycrystalline, one with amorphous and one with copper indium diselenide (CIS) solar cells. All modules have only series-connected cells. MPP power and efficiency are given for standard testing conditions (see section on 'Further electrical solar cell parameters', p136). The module efficiency is always lower than the theoretical cell efficiency since the module contains inactive areas between the cells. The module efficiency of the BP module is 13.5 per cent, whereas the efficiency of a single cell is 16 per cent. Almost all modules only have two bypass diodes; only a few modules contain a higher number of bypass diodes.

Table 4.6 *Technical Data for Various Photovoltaic Modules*

Name		SM 55	BP585	NT51A85E50-ALF	UPM 880	ST40
Manufacturer		Siemens	BP Solar	Sharp	ASE	Unisolar	Siemens
Number of cells		36 (3·12)	36 (4·9)	36 (4·9)	36 (4·9)	–	–
Cell type		mono-Si	mono-Si	mono-Si	poly-Si	a-Si	CIS
MPP power P_{MPP}	Wp	55	85	85.5	50	22	38
Rated current I_{MPP}	A	3.15	4.72	4.91	2.9	1.4	2.29
Rated voltage V_{MPP}	V	17.4	18.0	17.4	17.2	15.6	16.6
Short circuit cur. I_{SC}	A	3.45	5.00	5.5	3.2	1.8	2.59
Open circuit volt. V_{OC}	V	21.7	22.03	22.0	20.7	22.0	22.2
Temp. coeff. α_{ISC}	%/°C	+0.04	+0.03	+0.05	+0.09	*	+0.01
Temp. coeff. α_{VOC}	%/°C	–0.34	–0.34	–0.35	–0.38	*	–0.60
Temp. coeff α_{PMPP}	%/°C	*	*	–0.53	–0.47	*	*
Module efficiency	%	12.9	13.5	13.4	11.5	5.4	8.9
Length	mm	1293	1188	1200	965	1194	1293
Width	mm	329	530	530	452	343	329
Weight	kg	5.5	7.5	8.5	6.1	3.6	7.0
Bypass diodes		2	2	36	2	13	1

Note: * = no information

SOLAR GENERATOR WITH LOAD

Resistive load

The previous sections described characteristics of solar cells, modules and generators only. In reality, photovoltaic modules should provide electricity that is used by the consumer with an electric load.

The simplest load is an *electric resistance R* (Figure 4.30). A straight line describes the resistance characteristic. Ohm's law describes the relation between current and voltage:

$$I = \frac{1}{R} \cdot V \qquad\qquad (4.67)$$

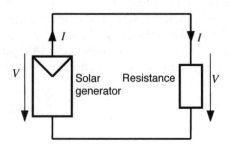

Figure 4.30 *Solar Generator with Resistive Load*

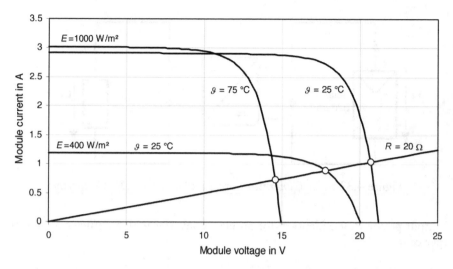

Figure 4.31 *Solar Module with Resistive Load at Different Operating Conditions*

If the current I through the resistance is set equal to the current of the solar cell [see Equation (4.34)], the common voltage and the operation point can be found by solving the equation for the voltage V. However, numerical methods are needed to obtain the solution.

For a *graphical estimation of the operating point*, I-V characteristics of the resistance and solar cell characteristics are drawn into the same diagram. The intersection of both characteristics then provides the operating point.

Figure 4.31 illustrates that the operating point of a solar module varies strongly with the operating conditions. Here, the module is operated close to the MPP at an irradiance of 400 W/m² and a temperature of 25°C. At other irradiances and temperatures, the module is operated sub-optimally and the output power is much less than the possible maximum power. Voltage and power at the resistive load vary significantly.

DC–DC converter

The power output of the solar module can be increased if a DC (direct current)–DC converter is connected between solar generator and load as shown in Figure 4.32.

The *converter* generates a voltage at the load that is different from that of the solar generator. Taking up the previous resistance example, Figure 4.33 shows that the power output of the module increases at higher irradiances if the solar generator is operated at a constant voltage. The power output can be increased even more if the solar generator voltage varies with temperature, i.e. if the voltage increases with falling temperatures.

Good DC–DC converters have efficiencies of more than 90 per cent. Only a small part of the generated power is dissipated as heat. Input power P_1 and

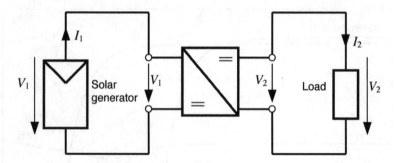

Figure 4.32 *Solar Generator with Load and DC–DC Converter*

output power P_2 are identical for an ideal converter with an efficiency of 100 per cent:

$$P_1 = V_1 \cdot I_1 = V_2 \cdot I_2 = P_2 \tag{4.68}$$

Since the voltages are different, the currents I_1 and I_2 are also different.

Buck converters

If the load voltage is always lower than the solar generator voltage, a so-called *buck converter*, as shown in Figure 4.34, can be used.

Switches and diodes are considered as ideal in the following calculations. The switch S is closed for the time period T_E and the current i_2 through the inductance L creates a magnetic field that stores energy. The voltage v_L at the inductance is:

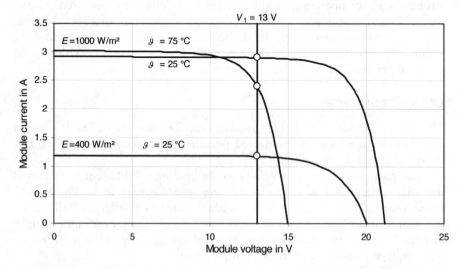

Figure 4.33 *Solar Module with Constant Voltage Load for Three Different Operating Conditions*

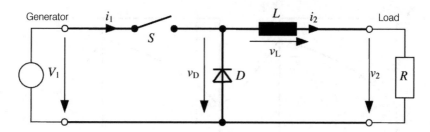

Figure 4.34 *Circuit of a Buck Converter with Resistive Load*

$$v_L = L \cdot \frac{di_2}{dt} \tag{4.69}$$

The switch is then opened for a time period T_A and the magnetic field of the inductance collapses and drives a current through the resistance R and the diode D. When neglecting the forward voltage drop at the diode, the output voltage v_2 at the contacts becomes:

$$v_2 = \begin{cases} v_D - v_L = V_1 - v_L & \text{with } v_L{>}0 \quad \text{for } 0 \le t \le T_E \\ v_D - v_L \approx -v_L & \text{with } v_L{<}0 \quad \text{for } T_E \le t \le T_E + T_A \end{cases} \tag{4.70}$$

After the period $T_S = T_E + T_A$ the procedure repeats. The mean voltage $\bar{v}_D$ with the *duty cycle* $\delta = T_E/T_S$ is:

$$\overline{v_D} = V_1 \cdot \frac{T_E}{T_S} = V_1 \cdot \delta \tag{4.71}$$

With $I_N = V_1/R$ and $\tau = L/R$, the current i_2 through inductance and load becomes:

$$i_2(t) = \begin{cases} I_N - (I_N - I_{min}) \cdot \exp(-t/\tau) & \text{for } 0 \le t \le T_E \\ I_{max} \cdot \exp(-(t - T_E)/\tau) & \text{for } T_E \le t \le T_S \end{cases} \tag{4.72}$$

The current i_2 is between the maximum current (Equation 4.73) and the minimum current (Equation 4.74):

$$I_{max} = I_N - (I_N - I_{min}) \cdot \exp(-T_E/\tau) = I_N \cdot \frac{1 - \exp(-T_E/\tau)}{1 - \exp(-T_S/\tau)} . \tag{4.73}$$

The minimum current is:

$$I_{min} = I_{max} \cdot \exp(-T_A/\tau) = I_N \cdot \frac{\exp(-T_A/\tau) - \exp(-T_S/\tau)}{1 - \exp(-T_S/\tau)} \tag{4.74}$$

Using I_{min} and I_{max} leads to the current i_2(Michel, 1992):

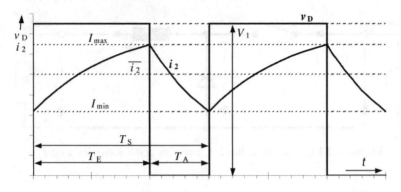

Figure 4.35 *Current i$_2$ and Voltage v$_D$ for a Buck Converter*

$$i_2(t) = \begin{cases} I_N + I_N \cdot \dfrac{\exp(-T_A / \tau) - 1}{1 - \exp(-T_S / \tau)} \cdot \exp(-t / \tau) & \text{for } 0 \leq t \leq T_E \\[2ex] I_N \cdot \dfrac{1 - \exp(-T_E / \tau)}{1 - \exp(-T_S / \tau)} \cdot \exp(-(t - T_E)/\tau) & \text{for } T_E \leq t \leq T_S \end{cases} \tag{4.75}$$

The average of the current $\overline{i}_2$ is:

$$\overline{i}_2 = I_N \cdot \frac{T_E}{T_S} = I_N \cdot \delta \tag{4.76}$$

In practice, the output voltage should be relatively constant. Therefore, the capacitors C_1 and C_2 shown in Figure 4.36 are inserted. Capacitor C_1 buffers the solar generator energy when the switch is open.

For an ideal inductance L, the mean output voltage is:

$$V_2 = \overline{v}_2 = V_1 \cdot \frac{T_E}{T_S} = V_1 \cdot \delta \tag{4.77}$$

with the input voltage V_1 and the duty cycle δ. Assuming ideal electronic components, the mean output current $\overline{i}_2$ is given by:

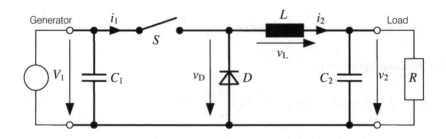

Figure 4.36 *Buck Converter with Capacitors*

$$I_2 = \overline{i_2} = \overline{i_1} \cdot \frac{T_S}{T_E} = \overline{i_1} \cdot \frac{1}{\delta} \tag{4.78}$$

This is obtained from the mean input current and the reciprocal of the duty cycle. If the mean output current I_2 is lower than:

$$I_{2,\text{lim}} = \frac{T_S}{2 \cdot L} \cdot V_2 \cdot (1 - \frac{V_2}{V_1}) \tag{4.79}$$

the current through the inductance during the switch blocking phase decreases until it reaches zero. The diode will then block the current and the voltage at the inductance becomes zero. The voltage and current thus go into a discontinuous operation mode. A suitable sizing of components can avoid this. If the limiting lower current $I_{2,\text{lim}}$ is known, the inductance can then be sized by:

$$L = T_S \cdot \left(1 - \frac{V_2}{V_1}\right) \cdot \frac{V_2}{2 \cdot I_{2,\text{lim}}} \tag{4.80}$$

Switching frequencies $f = 1/T$ between 20 kHz and 200 kHz seems to be a good compromise. The output capacity C_2 is obtained by:

$$C_2 = \frac{T_S \cdot I_{2,\text{glim}}}{4 \cdot \Delta V_2} \tag{4.81}$$

using the ripple of the output voltage V_2, i.e. the maximum desired voltage fluctuation ΔV_2.

Semiconductor components such as power field-effect bipolar transistors, or thyristors, for higher powers are mainly used for the switch S. Some integrated circuits (IC) exist that can directly control the duty period. For small power applications, the transistors are already integrated into some ICs.

Boost converters

Boost converters are suitable for applications with higher output voltages than input voltages. The principle boost converter structure is similar to that of the buck converter, except that the diode, switch and inductance change positions. Figure 4.37 shows the boost converter circuit.

If the switch S is closed, a magnetic field is created in the inductance L with the voltage $v_L = V_1$ ($v_L > 0$). When the switch opens, the voltage $v_2 = V_1 - v_L$ ($v_L < 0$) is applied to the load. This voltage is higher than the input voltage V_1. In this case, the voltage drop at the diode has been neglected. When the switch closes, the capacitor C_2 retains the load voltage. The diode D avoids the discharging of the capacitor through the switch S.

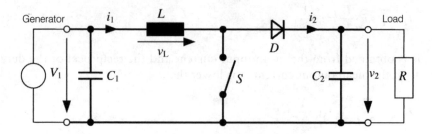

Figure 4.37 *Boost Converter Circuit*

The output voltage V_2 is:

$$V_2 = V_1 \cdot \frac{T_S}{T_A}$$ (4.82)

The dimensioning of L and C_2 with $I_{2,\text{lim}} = \frac{1}{2} \cdot V_1 \cdot (1 - V_1/V_2) \cdot T_S/L$ is possible:

$$L = U_1 \cdot (1 - \frac{V_1}{V_2}) \cdot \frac{T_S}{2 \cdot I_{2,\text{lim}}}$$ (4.83)

$$C_2 = \frac{T \cdot I_{2,\text{lim}}}{\Delta V_2}$$ (4.84)

Other DC–DC converters

Besides buck and boost converters, some other types of DC–DC converter exist. The output voltage of the *buck–boost converter* of Figure 4.38 is:

$$V_2 = -V_1 \cdot \frac{T_E}{T_A}$$ (4.85)

The *flyback converter* shown in Figure 4.39 uses a transformer instead of the inductance. The output voltage can be calculated in the same way as for the

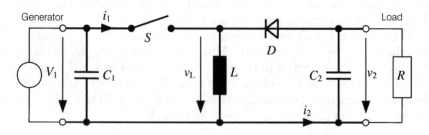

Figure 4.38 *Buck–Boost Converter Circuit*

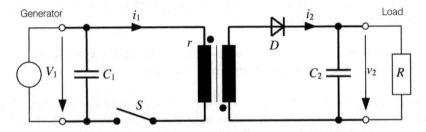

Figure 4.39 *Flyback Converter Circuit*

buck–boost converter, except that the transformation ratio r must be considered. This is defined by the ratio of the number of turns in the winding on either side of the transformer. Hence, the output voltage becomes:

$$V_2 = V_1 \cdot \frac{T_E}{T_A} \cdot \frac{1}{r} \qquad (4.86)$$

For high power applications a *push–pull converter* is used that needs more than one switch. If capacitors replace the inductances, a converter using the charge pump principle can be realized.

MPP tracker

The above-described voltage converters can maintain different voltages at the solar generator and at the load. If the solar generator voltage is set to a fixed value with a chosen duty cycle (see Figure 4.33), the energy yield is much higher than with a simple resistive load. On the other hand, the optimal operating voltage varies depending on irradiance and temperature. Therefore, a variation in the duty cycle of the DC–DC converter changes the solar generator voltage and thus can improve the energy yield.

Fluctuations in temperature ϑ have the highest influence on the optimal solar generator voltage. A temperature sensor attached to the back of a solar module can measure its temperature. With the temperature coefficient of the open circuit voltage (e.g. $\alpha_{VOC} = -3 \cdot 10^{-3}/°C$ to $-5 \cdot 10^{-3}/°C$ for silicon solar cells) the duty cycle for a buck converter can be estimated with the MPP voltage V_{MPP} at a reference temperature and the known output voltage V_2:

$$\delta = \frac{V_2}{V_1} = \frac{V_2}{V_{MPP}(\vartheta)} = \frac{V_2}{V_{MPP(\vartheta=25°C)} \cdot (1 + \alpha_{VOC} \cdot (\vartheta - 25°C))} \qquad (4.87)$$

If the duty cycle additionally is adapted to the solar irradiance, the solar generator can be operated at the maximum power point in most cases. If the DC–DC converter operates the solar generator at its MPP, the converter is called an MPP tracker.

There are several methods for implementing MPP tracker control that will be described in the following section:

Sensor-controlled regulator: As described above, the MPP voltage is calculated as a function of the temperature and irradiance sensor input.

Control using a reference cell: The characteristics, i.e. open circuit voltage V_{OC} and short circuit current I_{SC}, of a solar cell mounted near the solar generator is recorded. These measurements allow the estimation of the MPP voltage V_{MPP}. Using the equation of the simple equivalent circuit, the MPP current I_{MPP} becomes:

$$I_{MPP} = I(V_{MPP}) = I_{SC} - I_S \cdot \left(\exp\left(\frac{V_{MPP}}{m \cdot V_T} \right) - 1 \right)$$
(4.88)

The derivative of the power with respect to the voltage is equal to zero at the power maximum:

$$\frac{d\,P(V_{MPP})}{dV} = \frac{d(V_{MPP} \cdot I(V_{MPP}))}{dV} = I(V_{MPP}) + V_{MPP} \cdot \frac{d\,I(V_{MPP})}{dV} = 0$$
(4.89)

Inserting I_{MPP} into this equation and solving it for the MPP voltage V_{MPP} results in:

$$V_{MPP} = m \cdot V_T \cdot \ln\left(\frac{I_{SC} + I_S}{I_S} \right) - m \cdot V_T \cdot \ln\left(1 + \frac{V_{MPP}}{m \cdot V_T} \right) = V_{OC} - m \cdot V_T \cdot \ln\left(1 + \frac{V_{MPP}}{m \cdot V_T} \right)$$
(4.90)

Numerical or approximation methods can be used to solve this equation. The MPP voltage can be obtained from the open circuit voltage. A more precise estimate can be made using the two-diode model:

Oscillating search control (hill climbing): Voltage and current are measured at the converter input or output and the power is calculated and stored.

Figure 4.40 shows the principle of this system. Small changes in the duty cycle cause a voltage change. The power is then estimated again. If the power increases, the duty cycle is changed again in the same direction. Otherwise, the duty cycle is changed in the opposite direction. For constant output voltages, the search for the maximum output current is sufficient. In this case, the power itself need not be estimated.

Zero transit method: Generator voltage and current are measured and multiplied. A derivative unit estimates the derivative dP/dV. The generator voltage is increased or decreased depending whether the derivative is positive or negative.

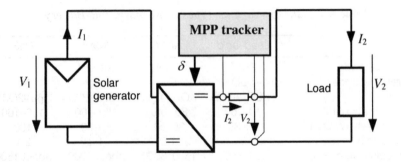

Figure 4.40 *Structure of MPP Trackers*

Control with differential changes: Voltage and current are measured and their differential change estimated. With

$$\frac{\mathrm{d}P}{\mathrm{d}V} = \frac{\mathrm{d}(V \cdot I)}{\mathrm{d}V} = I + V \cdot \frac{\mathrm{d}I}{\mathrm{d}V} = 0 \ , \text{ we get}$$

$$I \cdot \mathrm{d}V = -V \cdot \mathrm{d}I \tag{4.91}$$

i.e. the electronics must balance both quantities.

Control with characteristics method: This method also measures voltage and current. Starting with the open circuit voltage $V_A = V_{OC}$, voltage and current are changed alternately in the following manner:

$$V_B = k \cdot V_A \ \text{ and } \ I_C = k \cdot I_B \tag{$k < 1$}$$

After a few steps, this method obtains two operating points, one to the left and one to the right of the MPP; the controller oscillates between these points.

Various MPP trackers have difficulty in finding the optimal operating point when the solar generator is partially shaded, in rapidly changing conditions or for non-standard modules. Shading occurring over prolonged periods of time can cause high losses; therefore, a good MPP tracker should also provide good results for irregular operating conditions, in which multiple local maxima can occur in the I-V characteristic (see Figure 4.28, p146). When an unusually low generator power indicates a shading situation, the MPP tracker must scan through the whole voltage range to find the maximum for the optimal power output.

ELECTRICITY STORAGE

Types of batteries

Consumers are hardly ever connected directly to a solar generator. In reality, photovoltaic systems are more complex. In the absence of any kind of storage,

Table 4.7 *Data for Various Types of Rechargeable Battery*

	Lead–acid	NiCd	NiMH	NaS
Positive electrode	PbO_2	NiOOH	NiOOH	S
Negative electrode	PbO	Cd	metals	Na
Electrolyte	$H_2SO_4+H_2O$	$KOH+H_2O$	$KOH+H_2O$	β-Al2O3
Energy density (Wh/l)	10–100	80–140	100–160	150–160
Energy density (Wh/kg)	25–35	30–50	50–80	100
Cell voltage (V)	2	1.2	1.2	2.1
Charge/discharge cycles	500–1500	1500–3000	about 1000	about 1500
Operating temperature (°C)	0–55	–20 to 55	–20 to 45	290–350
Self-discharge rate (%/month)	5–15	20–30	20–50	0
Wh efficiency	70–85%	60–70%	60–85%	80–95%

there would be periods without any power for a solely photovoltaic-powered system, which clearly is not desirable.

Storage systems can be classified into short-term storage for a few hours or days to cover periods of bad weather and long-term storage over several months to compensate for seasonal variations in the solar irradiation in summer and winter. Since long-term storage is extremely expensive, the photovoltaic generator is usually oversized so that it also can provide sufficient energy in wintertime. Another solution is hybridization with wind or diesel generators.

Secondary electrochemical elements are mainly used for storage over short- and medium-term periods; they are usually called batteries. For economic reasons, the lead–acid battery dominates the current market. When higher energy densities are needed due to weight considerations, for example, in laptop computers, other batteries such as nickel–cadmium (NiCd) or nickel–metal hydride (NiMH) are used. Other batteries such as sodium–sulphur (NaS) have been tested for use in electrical (battery-powered) vehicles but are no longer being developed. Table 4.7 summarizes the data for various types of rechargeable battery.

Lead–acid battery

Today, the most common battery for electricity storage is the rechargeable lead–acid battery. The main reason is cost. The car industry, especially, prefers lead–acid batteries. So-called solar batteries have a slightly modified structure compared with car batteries and achieve longer lifetimes. However, the principle structure of the solar battery is similar to the car battery. It has two electrodes. In the charged state, the positive electrode consists of lead dioxide (PbO_2) and the negative electrode of pure lead (Pb). A membrane embedded in a plastic box separates the two electrodes. Diluted sulphuric acid (H_2SO_4) fills the empty space between the two electrodes. A fully charged lead–acid battery has an acid density of about 1.24 kg/litre at a temperature of 25°C, and this

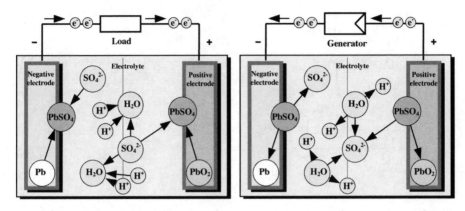

Figure 4.41 *Charging and Discharging a Lead–Acid Battery*

density changes with the temperature and charge state. An acid density meter or a voltmeter can indicate the charge state of a battery. Since the nominal voltage of one lead–acid battery cell is 2 V, six cells are connected in series to get the common operating voltage of 12 V. The number of cells can be adapted for other voltage levels.

Figure 4.41 illustrates the chemical reactions inside the lead–acid battery; they are represented by the following equations:

Negative electrode: $Pb + SO_4^{2-} \xleftrightarrow[\leftarrow\text{Charge}]{\text{Discharge}\rightarrow} PbSO_4 + 2e^-$

$$(4.92)$$

Positive electrode: $PbO_2 + SO_4^{2-} + 4H^+ + 2e^- \xleftrightarrow[\leftarrow\text{Charge}]{\text{Discharge}\rightarrow} PbSO_4 + 2H_2O$

$$(4.93)$$

Net reaction: $PbO_2 + Pb + 2H_2SO_4 \xleftrightarrow[\leftarrow\text{Charge}]{\text{Discharge}\rightarrow} 2PbSO_4 + 2H_2O \qquad (4.94)$

When discharging a lead–acid battery, the electrode material and the electrolyte react to form lead sulphate ($PbSO_4$). This reaction releases electrons that can be used as electrical energy by the consumer. Electrical energy must be fed into the battery for charging. The $PbSO_4$ at the electrodes transform to Pb and PbO_2 again. The charging process needs more energy than is set free during discharging. The charge efficiency is the ratio of the discharge over the charge. For the charge efficiency, a distinction is made between the *Ah efficiency* η_{Ah} and the *Wh efficiency* η_{Wh}. The Ah efficiency is calculated on the basis of integrated currents; the Wh efficiency considers currents and voltages during the discharge and charge periods needed to regain the same charge level as follows:

$$\eta_{Ah} = \frac{Q_{out}}{Q_{in}} = -\frac{\displaystyle\int_0^{t_{discharge}} I \cdot dt}{\displaystyle\int_0^{t_{charge}} I \cdot dt}$$

(4.95)

$$\eta_{Wh} = \frac{Q_{out} \cdot V_{out}}{Q_{in} \cdot V_{in}} = -\frac{\displaystyle\int_0^{t_{discharge}} V \cdot I \cdot dt}{\displaystyle\int_0^{t_{charge}} V \cdot I \cdot dt}$$

(4.96)

The Wh efficiency is always lower than the Ah efficiency because the battery voltage during charging is higher than during discharging. The Ah efficiency of a lead–acid battery is between 80 and 90 per cent depending on the battery type; the Wh efficiency is about 10 per cent lower.

Battery *self-discharge*, which causes additional losses, reduces the system efficiency. The self-discharge rate increases with the temperature and is about 0.3 per cent per day or 10 per cent per month at temperatures of 25°C. However, some battery types provide lower self-discharge rates.

The *usable capacity* of a battery depends on the discharge current, as shown in Figure 4.42. The capacity decreases with higher discharge currents and the end of discharge voltage is reached earlier.

To compare different rechargeable battery types, the capacity is often reported in combination with the discharge duration. C_{100} means that this capacity is usable when the discharge current is chosen so that the battery

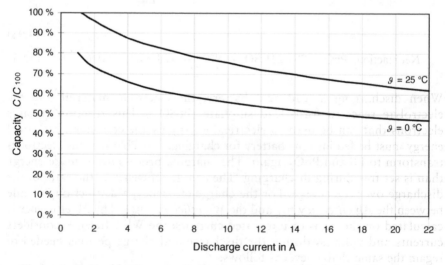

Figure 4.42 *Usable Capacity Related to C_{100} = 100 A h of a Lead–Acid Battery as a Function of the Discharge Current and Temperature*

Table 4.8 *Dependence of the Open Circuit Voltage and the Charge Density on the State of Charge of a 12-V Lead–Acid battery*

State of charge (SOC)	100%	75%	50%	25%	0%
Voltage in V	12.7	12.4	12.2	12.0	11.9
Acid density in kg/l	1.265	1.225	1.190	1.115	1.120

reaches the end of discharge voltage after 100 hours. A rechargeable battery with a capacity of C_{100} = 100 A h has a nominal discharge current of I_{100} = C_{100}/100 hours = 1 A. If the battery is discharged in 10 hours with a current of 8 A, the capacity C_{10} is reduced to less than 80 per cent of C_{100}. The usable capacity decreases to about 50 per cent at a temperature of 0°C and a discharge time of 5 hours. The lifetime of the battery, i.e. the number of cycles achievable, decreases with increasing temperature and discharge depth. The recommended maximum depth of discharge is normally about 80 per cent, whereas depths of discharge of more than 50 per cent (filling level below 50 per cent) should be avoided if possible.

Table 4.8 shows the open circuit voltage and acid density of a 12-V lead–acid battery. The nominal acid density for 100% charge varies between 1.22 kg/litre and 1.28 kg/litre depending on battery design and operational area. Higher acid densities improve the operating behaviour at low temperatures, whereas lower acid densities reduce corrosion. The given voltages are only valid for open circuit operation, i.e. the battery has not been charged or discharged recently. A temperature compensation of –25 mV/°C can be expected. When charging or discharging the battery, the voltage is above or below the open circuit voltage. The voltage difference compared with the open circuit voltage depends on the current. Figure 4.43 shows the voltage over the discharge time. Starting at the open circuit voltage of 12.7 V for a fully charged battery, the voltage falls depending on the discharge current. If the initial charge is lower, the voltage drop is higher.

The battery's state of health also influences the voltage slightly. The voltage is also an indicator of the *state of charge (SOC)* of a rechargeable battery (see Table 4.9).

The rechargeable battery should be protected against deep discharge or overcharging. If the battery is totally empty, crystalline lead sulphate is created. This type of lead sulphate is difficult to reconvert and some material will remain in the crystalline form. This damages the battery permanently. Therefore, deep discharge should be avoided in any case. This can be achieved in most cases by switching off the load at about 30 per cent of the remaining capacity. At common operating conditions, this is equivalent to a battery voltage of about 11.4 V. Lower voltages for ending discharge can be chosen for higher discharge currents above I_{10}. In addition, if the battery is not used for a long time, damage as a result of deep self-discharge is possible. The battery should be recharged from time to time to minimize the risk of damage.

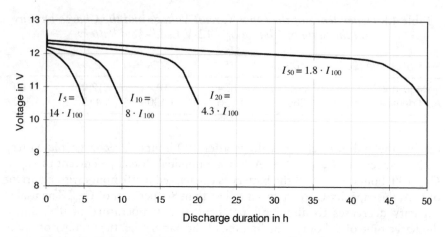

Figure 4.43 *Battery Voltage as a Function of Discharge Time and Discharge Current*

If the lead–acid battery is continuously charged, it starts to produce gas at a voltage of 14.4 V; the electrolysis decomposes the water within the electrolyte into hydrogen and oxygen and these gases escape from the battery. Therefore, the battery must be refilled with water from time to time. Continuous strong gassing can damage a battery. To protect the battery, charging should be stopped at voltages between 13.8 V and 14.4 V. However, it is advisable to charge lead–acid batteries until they begin gassing from time to time to mix the electrolyte thoroughly.

The batteries should be placed in a dry room at moderate temperatures. Battery gassing can produce explosive oxyhydrogen, so good *ventilation* of battery rooms is essential.

An equivalent circuit that describes battery behaviour is essential for exact simulations. However, an equivalent circuit that describes the battery perfectly is difficult to find, because many parameters such as temperature, current, state of charge (SOC) or state of health influence the operation. A standing joke is that a perfect model for a rechargeable battery is a black bucket with hole. However, several battery models have been developed over recent decades. One of these models, the Gretsch equivalent circuit, is described here as an example (Gretsch, 1978).

Table 4.9 *State of Charge Estimation for a 12-V Lead–Acid Battery Based on Measured Operating Voltages*

Voltage range (V)	State of charge (SOC)
>14.4	Stop charging, battery is full
13.5–14.1	Normal voltage range during charging without load
12.0–14.1	Normal voltage range during charging with load
11.5–12.7	Normal voltage range during discharging
11.4	Disconnect load, start charging

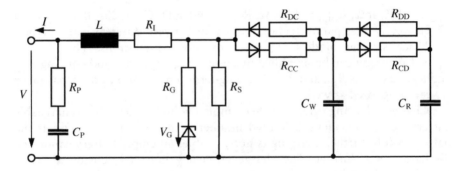

Source: Gretsch, 1978

Figure 4.44 *Gretsch Equivalent Circuit of a Lead–Acid Battery*

Figure 4.44 shows the *equivalent circuit* for one cell of a lead–acid battery. Most of the resistances of the equivalent circuit depend on temperature ϑ, state of charge SOC and the battery current I. The equation

$$R(\vartheta, p, I) = R_0 \cdot r(\vartheta) \cdot r\,(SOC) \cdot r(I) \tag{4.97}$$

considers these influences. R_0 is a constant base resistance that is modified by the three factors r depending on the operating conditions.

Table 4.10 describes the different elements of the equivalent circuit. Gretsch has found the parameters L = 1–10 μH, R_{I0} = 50 mΩ, R_P = 1.4 Ω, C_P = 1 μF, R_{G0} = 765 mΩ, V_G = 2.4 V, R_S = 5–10 kW, R_{DC0} = 25 mΩ,

Table 4.10 *Elements of the Lead–Acid Battery Equivalent Circuit*

Symbol	Component	Chemical–physical description
L	Inductance	Filamentary current separation across plates
R_I	Transient internal resistance	Metallic and ionic conductance
R_P	Polarization resistance	Re-orientation losses of dissociation product
C_P	Polarization capacity	Displacement current without chemical conversion
R_G	Gassing resistance	Inhibition of water decomposition
V_G	Gassing voltage	Start of water decomposition
R_S	Self-discharge resistance	Minor reaction, soiling
R_{DC}	Discharge conversion resistance	Inhibition of forming $PbSO_4$
R_{CC}	Charge conversion resistance	Inhibition of dissolving $PbSO_4$
R_{DD}	Discharge diffusion resistance	Acid concentration gradient
R_{CD}	Charge diffusion resistance	Acid concentration gradient
C_W	Work capacity	Direct convertible acid volume in pores
C_R	Rest capacity	Acid volume between plates, convertible mass

R_{CC0} = 140 mΩ, R_{DD0} = 40 mΩ, R_{CD0} = 40 mΩ, C_A = 20–50 F, C_R = 20 kF. However, these parameters can vary a lot between different battery types.

Much simpler models are used for most simulations. The Gretsch equivalent circuit can be simplified by ignoring C_P, R_P and L and unifying R_{DC} and R_{CC} as well as R_{DD} and R_{CD}. For long-term considerations, the capacitance C_A can be ignored as well.

Charge balancing offers another simple method to describe rechargeable battery behaviour. Here, a limited number of features is sufficient. If the battery is full, further charging is not possible; an empty battery cannot be discharged any further. The charge efficiency must be considered when charging and discharging a battery. Finally, self-discharge losses must be considered. Most simulation programs use simple charge balancing methods to get fast and adequate results.

Other rechargeable batteries

Other more expensive rechargeable battery types such as NiCd or NiMH are used in addition to the lead–acid battery. They have the advantages of higher energy density, fast charging capability and longer lifetime.

Nickel–cadmium (NiCd) *batteries* have the following advantages compared with lead–acid batteries:

- higher cycle number
- larger temperature range
- possibility of higher charge and discharge currents
- fewer problems with deep discharge.

On the other hand, NiCd batteries have the disadvantages of higher costs and the so-called *memory effect*. If charging of a NiCd battery is stopped before the full capacity is reached, the capacity decreases. Repeated full charging and discharging partly counteracts the capacity reduction; however, the memory effect is one of the most important problems for this type of battery.

Materials used in the production of NiCd batteries are the metals nickel and cadmium. The electrolyte is diluted potash lye with a density of between 1.24 kg/litre and 1.34 kg/litre. The chemical reactions in NiCd batteries are:

Negative electrode: $Cd + 2OH^- \xrightarrow[\leftarrow \text{Charge}]{\text{Discharge} \rightarrow} Cd(OH)_2 + 2e^-$ (4.98)

Positive electrode: $2NiO(OH) + 2H_2O + 2e^- \xrightarrow[\leftarrow \text{Charge}]{\text{Discharge} \rightarrow} 2Ni(OH)_2 + 2OH^-$

(4.99)

Net reaction: $2NiO(OH) + Cd + 2H_2O \xrightarrow[\leftarrow \text{Charge}]{\text{Discharge} \rightarrow} 2Ni(OH)_2 + Cd(OH)_2$

(4.100)

The nominal voltage of a NiCd cell of 1.2 V is lower than that of a lead–acid battery cell. NiCd batteries are mainly used as household batteries as well as for laptops or electric cars.

One major disadvantage of NiCd batteries is the use of environmentally problematic materials. It surely cannot be avoided that constituent materials of disposed batteries are released into the environment after the end of the battery's useful life. Cadmium accumulates in the food chain, and in human bodies, because it is excreted only partially. High cadmium contamination can cause organ damage or cancer.

Nickel–metal hydride (NiMH) batteries are much less environmentally problematic. Applicable metals are nickel, titanium, vanadium, zirconium or chrome alloys. However, small amounts of toxic materials are also used for these batteries. The electrolyte is diluted potash lye, the same as for NiCd batteries. Besides good environmental compatibility, NiMH batteries have further advantages compared to NiCd batteries such as higher energy density and the absence of the memory effect. Disadvantages are the smaller temperature range and the high self-discharge rate (about 1 per cent per day). Since the cell voltage of 1.2 V is the same as for NiCd batteries, NiMH batteries can easily replace NiCd batteries.

The chemical reactions in NiMH batteries are:

Negative electrode: $MH + OH^- \underset{\leftarrow Charge}{\overset{Discharge \rightarrow}{\longleftrightarrow}} M + H_2O + e^-$ (4.101)

Positive electrode: $NiOOH + H_2O + e^- \underset{\leftarrow Charge}{\overset{Discharge \rightarrow}{\longleftrightarrow}} Ni(OH)_2 + OH^-$ (4.102)

Net reaction: $NiOOH + MH \underset{\leftarrow Charge}{\overset{Discharge \rightarrow}{\longleftrightarrow}} Ni(OH)_2 + M$ (4.103)

Estimation of the state of charge for NiCd and NiMH batteries is more complicated compared with lead–acid batteries. The temperature influence is greater and the voltage of a fully charged NiCd or NiMH battery even decreases a little.

Other rechargeable battery types such as sodium–sulphur (*NaS*) batteries promise advantages of higher energy densities; however, problems with high operating temperatures and dangerous materials such as sodium have not yet been resolved fully. Because only prototypes of these batteries exist, they are not discussed in detail.

Battery systems

The simplest battery system consists only of a photovoltaic generator, a battery and a load. Since the internal resistance of the photovoltaic generator is very small, the battery discharges through the photovoltaic generator if the solar irradiance is low. A *blocking diode* between the photovoltaic generator and the battery, as shown in Figure 4.45, can avoid these reverse currents from the battery to the photovoltaic generator; however, this diode causes permanent losses:

$$P_{L,\text{diode}} = I_{PV} \cdot V_D \tag{4.104}$$

Therefore, diodes with low forward voltages such as Schottky diodes ($V_D \approx$ 0.55 V) are often used. Cables cause further losses; a connection cable with cross-section A, specific resistance ρ and cable lengths l_1 and l_2 for the cables from the photovoltaic generator to the battery and back, respectively, causes the following losses:

$$P_{L,\text{cable}} = I_{PV} \cdot (V_{C1} + V_{C2}) = I_{PV}^2 \cdot (R_{C1} + R_{C2}) = I_{PV}^2 \cdot \frac{\rho}{A} \cdot (l_1 + l_2) \tag{4.105}$$

A copper cable (ρ_{Cu} = 0.0175 Ω mm²/m) with cable lengths ($l_1 = l_2$ = 10 m), a cross-section of A = 1.5 mm² and a current I_{PV} = 6 A causes cable losses of $P_{L,\text{cable}}$ = 8.4 W. Assuming a photovoltaic generator power of 100 W, these cable losses plus the blocking diode losses of 3.3 W are considerable at 12 per cent of the power generated. To minimize losses, cables should be as short as possible and the cable cross-section appropriately large. For a 12-V battery system, a voltage drop of 3 per cent, or 0.35 V, is acceptable in the cables from the photovoltaic generator to the battery and 7 per cent, or 0.85 V, from the battery to the load. For the above example, the cable cross-section must therefore be 6 mm².

For systems with higher power, the losses can be reduced if some batteries are connected in series. This increases the battery voltage V_{Bat} and decreases the current flow and thus the losses.

The photovoltaic generator has the voltage:

$$V_{PV} = V_{Bat} + V_D + V_{C1} + V_{C2} \tag{4.106}$$

The diode voltage V_D is nearly constant, the cable voltage drop V_{C1} and V_{C2} are proportional to the photovoltaic current I_{PV}. The battery voltage V_{Bat} depends on the charge current and state of charge. Hence, the voltage at the photovoltaic generator increases slightly with rising currents and increasing

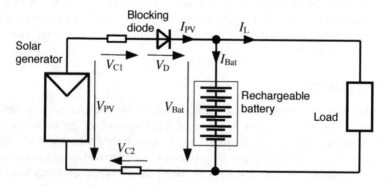

Figure 4.45 *Simple Photovoltaic System with Battery Storage*

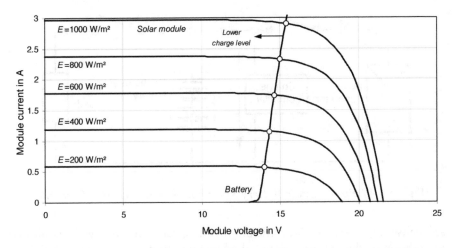

Figure 4.46 *Operating Points of a Solar Module Connected to Battery Storage with a Blocking Diode and 0.1 Ω Cable Resistance without Load*

irradiances, and it varies with the battery state of charge. If a photovoltaic generator is connected directly to a battery, a good operating point is achieved for a wide irradiance range (see Figure 4.46). Therefore, DC–DC converters and MPP trackers are very rarely used in battery systems. In some cases, the power consumption of the additionally required electronics is higher than the possible energy gain. Only for inhomogenously irradiated solar generators do MPP trackers have some advantages.

Rechargeable batteries in simple battery systems with the photovoltaic generator and load directly connected to the battery are not protected against deep discharge or overcharging. Such a simple system should be chosen only if negative operating conditions for the battery can be avoided with certainty; otherwise, the battery can be damaged very quickly.

Therefore, most battery systems use a charge controller (see Figures 4.47 and 4.48). Charge-controller lead–acid battery systems usually work on the basis of voltage control. The charge controller measures the battery voltage V_{Bat}. If it falls below the deep discharge voltage (11.4 V for a 12-V lead–acid battery), the switch S_2 disconnects the load from the battery. When the battery is charged again, i.e. the battery voltage rises above an upper threshold, the switch reconnects the load. If the battery voltage rises above the end of charge voltage (about 14.4 V for a 12-V lead–acid battery), the switch S_1 stops charging. The *series charge controller* (see Figure 4.47) and the *parallel or shunt charge controller* are two principal charge controller types (see Figure 4.48).

Power semiconductors such as power field-effect transistors (power MOSFETs) are normally used as switches. Continuous forward losses at the switch S_1 are a disadvantage of the series charge controller. The forward resistance of good MOSFETS is less than 0.1 Ω; nevertheless, for a current of 6 A, the field-effect transistor BUZ 11 with a forward resistance of 0.04 Ω

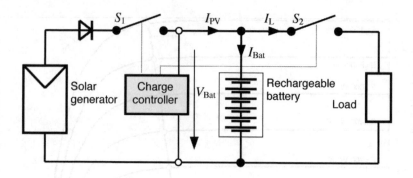

Figure 4.47 *Photovoltaic Battery System with Series Charge Controller*

still causes losses of 1.44 W. If the photovoltaic generator voltage is monitored in addition to the battery voltage, the blocking diode can be omitted and the forward losses reduced. In this case, the charge controller must open the switch S_1 if the solar generator voltage falls below the battery voltage.

The parallel charge controller is the most commonly used type of charge controller. If the battery is fully charged, the charge controller short-circuits the solar generator across switch S_1. The solar generator voltage falls to the voltage drop across the switch (<1 V). The blocking diode avoids reverse currents from the battery across the switch. If the generator is in the regular mode of operation, the short circuit does not cause any problems; however, if the solar generator is partially shaded and thus not irradiated homogenously, the short circuit conditions can strain the shaded cells very significantly.

Individual battery cells of large battery systems can be severely strained due to small differences between the cells. These cells age faster, and if one cell fails, other cells will be affected. Therefore, battery management systems that not only monitor the voltage of the whole battery unit but also the voltage of single cells should be used for large battery systems for optimal battery protection.

Besides these relatively simple battery systems with a photovoltaic generator, a battery, charge controller and a load, more complex systems exist.

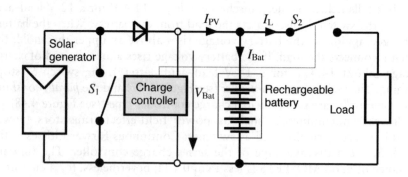

Figure 4.48 *Photovoltaic Battery System with Parallel Charge Controller*

Table 4.11 *Energetic Data for Hydrogen in its Normal State*
(ρ = 0.101 MPa, T = 273.15 K = 0°C)

Lower calorific value (LCV)	Upper calorific value (UCV)	Density
3.00 kWh/m_n^3	3.55 kWh/m_n^3	0.09 kg/m^3 (gaseous)
33.33 kWh/kg	39.41 kWh/kg	70.9 kg/m^3 (liquid, −252°C)

Note: 1 m_n^3 = 1 nominal cubic metre, equal to 0.09 kg of hydrogen gas

Battery backup systems can be connected to the grid with a photovoltaic inverter. Photovoltaic generators can be combined with a wind generator or diesel generator to create a hybrid system. This can reduce energy costs and increase system availability; however, operating complex systems needs a complex energy management system. If an AC load should be operated, an inverter must be integrated into the system; inverters for photovoltaic systems are described below.

Hydrogen storage and fuel cells

A promising option for future storage of large amounts of energy is hydrogen (H_2) storage (see energetic data in Table 4.11).

Electrolysis processes can produce hydrogen by using electricity as the driving force. In alkaline electrolysis, water is split into hydrogen and oxygen using two electrodes in a dilute alkaline electrolyte (see Figure 4.49).

The following equations describe the reactions:

$$\text{Cathode: } 2H_2O + 2e^- \rightarrow H_2 + 2OH^- \tag{4.107}$$

$$\text{Anode: } 2OH^- \rightarrow \tfrac{1}{2}O_2 + H_2O + 2e^- \tag{4.108}$$

Alkaline electrolysis currently achieves efficiencies of 85 per cent. Besides alkaline electrolysis, other methods for generating hydrogen from water exist (e.g. membrane electrolysis or high-temperature vapour electrolysis). A smaller

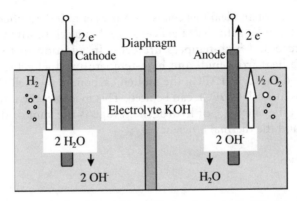

Figure 4.49 *Principle of Hydrogen Electrolysis with Alkaline Electrolyte*

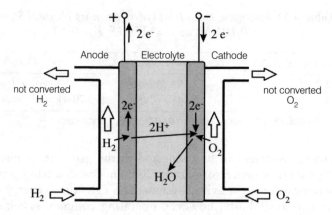

Figure 4.50 *Principle of the Fuel Cell with Acid Electrolyte*

amount of electric energy is sufficient to keep the process running at temperatures above 700°C; prototypes of pressurized electrolysers have already reached efficiencies of 90 per cent.

Gas and steam turbine power plants or fuel cells can generate electricity from hydrogen. However, *fuel cells* are the most promising option to date. A fuel cell with an acid electrolyte reverses the electrolysis by regenerating water (H_2O) from oxygen O_2 and hydrogen H_2 (see Figure 4.50); electrical energy is a result of this reaction. The anode collects electrons that are released. The hydrogen ions H^+ diffuse through the electrolyte to be collected at the cathode. There they coalesce with the oxygen ions and electrons from outside to form water molecules.

The equations of the reactions in the fuel cell are:

Anode: $H_2 \rightarrow 2H^+ + 2e^-$ (4.109)

Cathode: $\frac{1}{2}O_2 + 2H^+ + 2e^- \rightarrow H_2O$ (4.110)

To increase the voltage, several cells are connected in series into stacks (Figure 4.51).

The efficiency of an ideal fuel cell is 94.5 per cent. Today, efficiencies of up to 80 per cent can be achieved. Fuel cells are divided into low-temperature, mid-temperature and high-temperature cells. In addition to the electricity generated, the heat from mid- and high-temperature fuel cells, such as the phosphoric acid fuel cell (PAFC), the molten carbonate fuel cell (MCFC) and the solid oxide fuel cell (SOFC), can be used. This increases the total efficiency.

The total efficiency η_{tot} of hydrogen storage is the sum of the electrical efficiency η_{el} and the thermal efficiency η_{th}:

$$\eta_{tot} = \eta_{el} + \eta_{th}$$ (4.111)

Figure 4.51 *Photograph of a Fuel Cell Stack Prototype*

The electrical efficiency

$$\eta_{el} = \eta_{Ely} \cdot (1 - f_{Tr}) \cdot (1 - f_{St}) \cdot \eta_{FC,el} \tag{4.112}$$

is obtained from the electrolysis efficiency η_{Ely}, transport losses f_{Tr}, storage losses f_{St} and the electrical efficiency of the fuel cell $\eta_{FC,el}$. The electrical efficiency η_{el} of further-improved systems could be about 50 per cent. Assuming a thermal efficiency η_{th} of 20 per cent provides a total efficiency η_{tot} of 70 per cent.

Owing to the relatively poor electrical efficiency of the hydrogen storage chain, the direct use of electricity is preferred, even if the electricity must be transported over very large distances.

Other storage concepts

There are other possibilities for storing electrical energy apart from electrochemical storage in batteries or hydrogen storage. However, most of the other storage concepts are rarely used today. Other methods, such as pumped-storage plants, can be only used for very large applications. Other storage concepts are:

- capacitor banks
- storage in superconducting coils
- flywheel storage
- pumped-storage hydroelectric power plants
- compressed air storage.

INVERTERS

Inverter technology

It was assumed until now that loads are operated by direct current (DC); however, most consumer applications use alternating current (AC). An island grid inverter can operate AC devices with a photovoltaic stand-alone system. If the system is to be connected to a national grid, a grid-connected inverter is used. Both inverter categories are similar, but there are some differences in the details.

Power electronic devices are used today to convert DC to AC. Different types of semiconductor elements that can switch voltages higher than 1000 V or even currents higher than 1000 A are:

- power MOSFET (power field effect transistors)
- bipolar power transistors
- insulated gate bipolar transistors (IGBT)
- thyristors (controllable diodes)
- triacs (two-direction thyristors)
- gate turn-off (GTO) thyristors (switchable thyristors).

A short look at the semiconductor thyristor will cover the functionality of power electronic switches. Figure 4.52 shows the thyristor symbol with three contacts: A (anode), C (cathode) and G (gate).

If the control current i_G is equal to zero, the thyristor blocks at negative and positive voltages v. A control current i_G exceeding a positive maximum voltage v or, exceeding a maximum voltage increase dv/dt, switches the thyristor into forward mode. Then, a current i can flow through the thyristor if there is a positive voltage v. Normally, this is caused by a control current i_G. While the forward current i is above the holding current i_H, the thyristor remains conductive. A triac can be operated in both directions and a GTO thyristor can be turned off by a negative control current.

An inverter must periodically direct the current from one branch to another; this is called *commutation*. Therefore, different thyristors must switch alternately. If the energy needed to change the state of the thyristor comes from outside, for instance from the grid, the commutation is called external or line

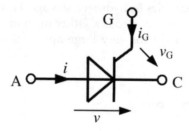

Figure 4.52 *Thyristor Symbol*

commutation. If the circuit can provide this energy itself, it is called self-commutation; however, such a circuit needs energy storage. An externally commutated inverter is not suitable for stand-alone operation. The grid defines the switching points of externally commutated inverters, whereas a self-commutated inverter must determine them itself. The voltage of grid-connected inverters must be synchronized with the grid voltage.

Besides the type of commutation, inverter technologies can be classified as:

- square-wave or trapezium inverters
- stair-step inverters
- pulse-width modulated or resonance inverters.

Other inverter differences depend on the operating mode. Grid-connected inverters have to fulfil stringent criteria to maintain high power quality. Therefore, amplitude, frequency and current shape must follow the rules of the grid operators. For grid-protection, the inverter must switch off immediately if the grid fails. However, island inverters do not have to meet these strict criteria.

Square-wave inverter

A very simple inverter circuit is the *two-pulse bridge connection (B2 connection)* shown in Figure 4.53. It consists of four thyristors. A transformer connects the circuit to the grid.

Thyristors 1 and 3 work together, as do thyristors 2 and 4. If these two groups switch periodically, they generate a square-wave alternating current at the transformer. Thyristors 1 and 2 can be replaced by non-controllable diodes for simplification; in this case, only half of the bridge must be controlled. This connection is then called a half-controlled bridge connection. The switching of the thyristors is delayed by the control angle α compared to the voltage zero crossing. Figure 4.54 shows the current of a B2 connection.

This shape differs significantly from that of a sinusoidal wave. To assess the current quality, Fourier analysis or *harmonic analysis* is used.

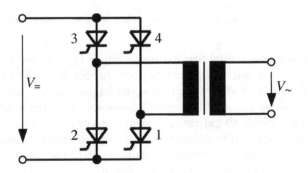

Figure 4.53 *Two-pulse Bridge Connection (B2)*

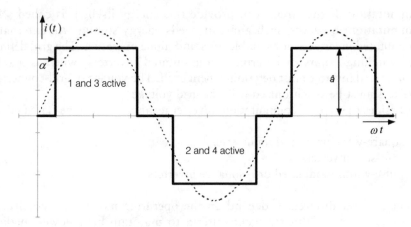

Figure 4.54 *Idealized Current of a Half-controlled B2 Bridge Connection*

A 2π periodical function $f(\omega t)$ in the interval $-\pi \leqslant \omega t \leqslant +\pi$ can be decomposed in a convergent series:

$$f(\omega t) = \frac{a_0}{2} + \sum_{n=1}^{\infty} \left(a_n \cdot \cos(n \cdot \omega t) + b_n \cdot \sin(n \cdot \omega t)\right) \tag{4.113}$$

The coefficients a_n and b_n can be estimated as:

$$a_n = \frac{1}{\pi} \cdot \int^{\pi} f(\omega t) \cdot \cos(n \cdot \omega t) \, d\omega t \quad (n = 0, 1, 2, ...) \tag{4.114}$$

$$b_n = \frac{1}{\pi} \cdot \int^{\pi} f(\omega t) \cdot \sin(n \cdot \omega t) \, d\omega t \quad (n = 1, 2, ...). \tag{4.115}$$

For the square-wave current of the B2 bridge connection with amplitude â and control angle α, harmonic analysis results in:

$$f(\omega t) = \frac{4 \cdot \hat{a}}{\pi} \cdot \left[\cos\alpha \cdot \sin\omega t + \tfrac{1}{3} \cdot \cos 3\alpha \cdot \sin 3\omega t + \tfrac{1}{5} \cdot \cos 5\alpha \cdot \sin 5\omega t + \cdots\right]. \tag{4.116}$$

Besides the desired sinusoidal first harmonic (order 1), this function consists of various oscillations with different periods (order $\geq$ 2). They are called harmonics. Figure 4.55 shows the first seven harmonics The square-shape is already recognizable when adding these harmonics. A perfect reproduction of the square shape needs all existing harmonics. In practice, harmonic analysis can be carried out with an oscilloscope and computer-based analysis. Proprietary devices can estimate the harmonic content independently.

According to harmonic analysis, the current $i(t)$ is composed of the first harmonic with the amplitude $\hat{i}_1$ and further harmonics with the amplitudes $\hat{i}_2$, $\hat{i}_3$ and so on:

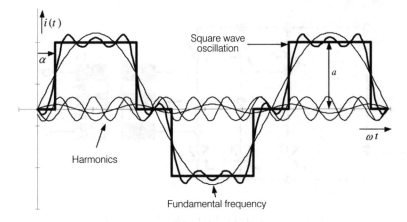

Figure 4.55 *Construction of a Square-wave Oscillation from Different Sinusoidal Harmonics*

$$i(t) = \hat{i}_1 \cdot \sin(\omega t) + \hat{i}_2 \cdot \sin(2 \cdot \omega t) + \hat{i}_3 \cdot \sin(3 \cdot \omega t) + \ldots \qquad (4.117)$$

Most national regulations or grid operator requirements lay down maxima for the current harmonics. Inverters can cause voltage harmonics in addition to current harmonics. These are also limited by maximal permissible values in relation to the first harmonic. The *harmonic distortion factor*

$$D = \sqrt{\sum_{n=2}^{40} \left(\frac{\hat{v}_n}{\hat{v}_1} \right)^2} \qquad (4.118)$$

which is calculated using the voltage harmonic amplitudes $\hat{v}_n$ and the first harmonic amplitude $\hat{v}_n$, describes the harmonic content of the voltage. The maximal permissible harmonic distortion factor is usually $D = 0.8$. Besides the amplitudes, root mean square (rms) values are often used for calculations. In Chapter 5, the section headed 'Alternating current (p205) describes the calculation of rms values in detail.

Another quantity of assessment is the *relative harmonic content*,

$$k = \sqrt{\frac{\sum\limits_{n=2}^{\infty} V_n^2}{\sum\limits_{n=1}^{\infty} V_n^2}} \qquad (4.119)$$

which is the ratio of all the rms harmonics except for the first one to all the rms harmonics. The relative harmonic content for good inverters is below 3 per cent. If the harmonic content is too high, filters must reduce it.

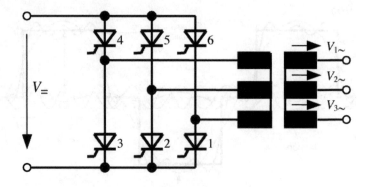

Figure 4.56 *Six-pulse Bridge Inverter (B6)*

A transformer is integrated into most inverters. This transformer separates the inverter from the grid and can transform the inverter voltage to a higher or lower grid voltage. However, a transformer always introduces losses and is not essential in all inverter designs. For transformerless inverters, more sophisticated protection methods are used because there is no longer any galvanic separation between the solar generator and the grid.

The described B2 bridge connection delivers an alternating current to a single phase of the grid. This can cause asymmetries in the three phases of the public grid. To overcome this, three-phase generation is preferred for inverter powers above 5 kW. Chapter 5 (wind power) describes the basics of three-phase currents. One circuit for the generation of three-phase current is the *six-pulse bridge connection* (B6 bridge), shown in Figure 4.56. The thyristors of this circuit switch in a way that generates three alternating currents and voltages shifted by 120°.

Besides the described B2 and B6 bridge connections, other circuits such as the M2 or M3 star connection or other bridge connections are used. However, the principle of these connections is similar to that of the above-described variants.

Pulse-width modulation

An inverter that works on the *pulse-width modulation* (PWM) principle also uses the above described B2 or B6 bridge circuits. However, the thyristors do not switch just once per half-wave: multiple switching generates pulses of different widths as shown in Figure 4.57. The sinusoidal fundamental wave is obtained after filtering. The quality of the sinusoidal oscillation is much better compared to rectangle inverters, i.e. PWM inverters have much lower harmonic content. Therefore, most inverters use the PWM principle today.

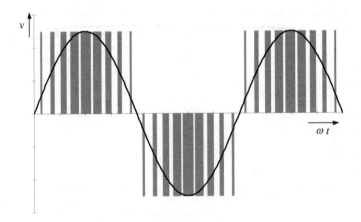

Figure 4.57 *Voltage using Pulse-width Modulation*

Photovoltaic inverters

Photovoltaic inverters differ from inverters for other applications. This is because they must operate the photovoltaic generator at the optimal operating point to generate the maximum power. Grid-connected inverters are therefore often combined with MPP trackers, i.e. DC–DC converters, to set a voltage at the generator that differs from the grid voltage. A battery charge controller is integrated into most stand-alone inverters.

Photovoltaic inverters operate at the rated power P_N for only very few hours in any year. Due to the changing solar irradiance, the inverters predominantly operate under part load. Therefore, it is very important that photovoltaic inverters have high efficiencies even when operating under part-load conditions. Furthermore, the inverter should never be oversized. For instance, if a 1-kW inverter operates with a 500-W_p photovoltaic generator, the input power reaches at most 50 per cent of the rated power, and losses due to permanent part-load operation may be very high. Furthermore, the self-consumption of a photovoltaic inverter should be minimal and the inverter should be switched off at night.

Figure 4.58 shows the *efficiency characteristics* over the range of relative input powers for two commercial inverters. The inverter with the higher rated power has a higher efficiency; however, both inverters show good part-load behaviour even for input power below 10 per cent of the rated value.

A representative efficiency is used to compare different inverters, the so-called *Euro efficiency* η_e, which is given as:

$$\eta_e = 0.03 \cdot \eta_{5\%} + 0.06 \cdot \eta_{10\%} + 0.13 \cdot \eta_{20\%} + 0.1 \cdot \eta_{30\%} + 0.48 \cdot \eta_{50\%} + 0.2 \cdot \eta_{100\%}$$

$$(4.120)$$

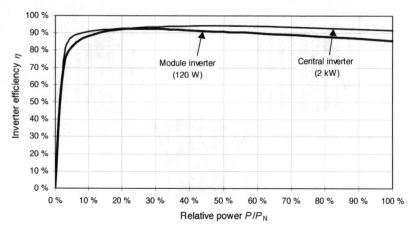

Figure 4.58 *Efficiency over a Range of Relative Photovoltaic Generator Powers*

It considers the typical part-load operation for irradiation regimes in central Europe.

The ideal energy yield E_{ideal} of a photovoltaic system with photovoltaic area A_{PV}, module efficiency η_{PV} and solar irradiation H_{Solar} is given by:

$$E_{ideal} = A_{PV} \cdot \eta_{PV} \cdot H_{Solar} \tag{4.121}$$

Assuming a 10 m² inclined area with an annual solar irradiation of 1100 kWh/m², modules with an efficiency of 10 per cent could generate 1100 kWh per year in the ideal case. In reality, the energy yield of the photovoltaic system is much lower. The average real efficiency of solar modules is lower than the nominal efficiency due to soiling, shading and elevated operating temperatures. Inverter losses reduce the yield further. The so-called *performance ratio PR* describes the ratio of the real and ideal energy:

$$E_{real} = PR \cdot E_{ideal} \tag{4.122}$$

Good systems have PRs of more than 0.75 = 75 per cent. This value can be used for designing a new system if no shading losses are to be considered. Very good systems can even reach PRs that are higher than 0.8. Problematic systems can have PRs below 0.6. Inverter failures, MPP tracking problems, module failure or solar generator shading are the main reasons for low PR values.

Figure 4.59 shows the connection of solar modules to a central inverter. A number of modules are connected in series to form a string until the desired voltage is reached. Several identical strings can be connected in parallel to increase the power of the generator. Blocking diodes that were often used in older systems can be omitted because their protection capability is low and they cause permanent forward losses.

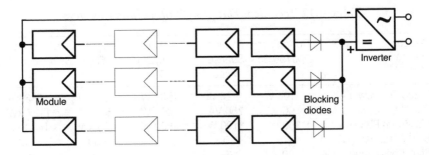

Figure 4.59 *Photovoltaic System with Parallel Strings and Central Inverter*

For partly shaded systems or modules with different power outputs, this connection scheme can cause relatively high losses. Therefore, string inverters with decoupled parallel strings can reduce the losses. Their nominal power is in the range of 1 kW. However, module inverters promise the best performance for partially shaded photovoltaic generators. They can set a different operating voltage for each module. The cabling is simpler because no DC cables are needed. Another advantage is the system modularity. However, the disadvantages of module inverters are lower nominal efficiencies and higher specific costs. Figure 4.60 shows both versions: string and module inverters.

Finally, Table 4.12 shows *technical data* for several inverters. It is apparent that the efficiency increases slightly with the rated power. The harmonic distortion of available inverters is between 1.5 and 5 per cent. Most of the inverters are designed using the PWM concept.

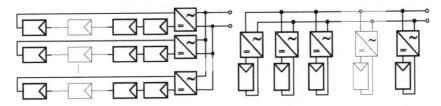

Figure 4.60 *Photovoltaic Generator with String Inverters (left) and Module Inverters (right)*

Table 4.12 *Technical Data for Photovoltaic Inverters*

Device	Dorfmüller DMI150/35	Fronius Sunrise Mirco	SMA SunnyBoy 2000	ACE 5001	Siemens Ss 4 x 300 kVA
Rated DC power	100 W	820 W	1900 W	50 kW	1086 kW
Rated AC power	90 VA	750 VA	1800 VA	50 kVA	1197 kVA
Maximum PV power	150 W	1100 W	2600 W	55 kW	1400 kW
Start of feed-in (W)	2.5	9	7	50	700
Stand-by demand (W)	0	5	7	35	40
Night consumption	0	0	0,1	0	30
DC MPP range (V)	28–50	120–300	125–500	300–420	460–730
Maximum distortion	3%	5%	4%	0.1%	act. Filter
Maximum efficiency (%)	89	92	96	97	97
Euro efficiency (%)	86.6	90.4	95.2	94.2	96.2

Source: data from Photon International, 2001

Chapter 5

Wind Power

INTRODUCTION

Wind energy is an indirect form of solar energy in contrast to the direct solar energy described in previous chapters. Solar irradiation causes temperature differences on Earth and these are the origin of winds. The wind itself can be used by technical systems. Wind can reach much higher *power densities* than solar irradiance: 10 kW/m^2 during a violent storm and over 25 kW/m^2 during a hurricane, compared with the maximum terrestrial solar irradiance of about 1 kW/m^2. However, a gentle breeze of 5 m/s (18 km/h, 11.2 mph) has a power density of only 0.075 kW/m^2.

The *history of wind power* goes back many centuries. Wind power was used for irrigation systems 3000 years ago. Historical sources give evidence for the use of wind power for grain milling in Afghanistan in the 7th century. These windmills were very simple systems with poor efficiencies compared to today's systems. In Europe, wind power became important from the 12th century onwards. Windmills were improved over the following centuries. Tens of thousands of windmills were used for land drainage in The Netherlands in the 17th and 18th centuries; these mills were sophisticated and could track the wind autonomously. In the 19th century numerous western windmills were used in North America for water pumping systems. Steam powered machines and internal combustion engines competed with wind power systems from the beginning of the 20th century. Finally, electrification made wind power totally redundant. The revival of wind power began with the oil crises of the 1970s. In contrast to the mechanical wind power systems of past centuries, modern wind converters almost exclusively generate electricity. Germany became the most advanced country for wind technology development in the 1990s. State of the art wind generators have reached a high technical standard and now have powers exceeding 4 MW. The German wind power industry alone has created more than 45,000 new jobs and has reached an annual turnover of more than €3500 million.

The high growth rate of the wind power industry indicates that wind power will reach a significant share of the electricity supply within the next two decades, and not only in Germany and Denmark (the other significant centre of development). Therefore, the main deciding factors are the legislative conditions. For instance, the Renewable Energy Sources Acts in Germany and Spain were the basic conditions for the wind power boom in these two countries. In most countries the potential for wind power utilization is enormous. Germany could provide one-third of its electricity demand and the UK could theoretically cover even more than its whole electricity demand with wind power.

Germany can be taken as example of the rapid development of wind power and its integration into the electricity supply structures. Most of the established utilities fear the competition and complain about the problems with line regulation that result from fluctuations in wind power; however, some utilities have demonstrated that improved wind speed forecasts can solve these problems. Even some environmental organizations protest against new wind installations. Their reasons are conservation, nature or noise protection; indeed, some of their arguments are justifiable. On the other hand, wind power is one of the most important technologies for stopping global warming. No doubt, wind generators change the landscape, but if we do not get global warming under control, coastal areas that would be protected by the reduction in global warming resulting from wind generator installation will most likely not exist far into the future.

The discussions of wind power make clear that a social consensus about future energy policy does not exist. It is hoped that this consensus can be achieved soon because the utilization of the enormous potential of the wind is essential for an economical, sustainable electricity supply.

THE WIND

Wind resources

The sun heats up air masses in the atmosphere. The spherical shape of the Earth, the Earth's rotation and seasonal and regional fluctuations of the solar irradiance cause spatial *air pressure differentials*. These are the source of air movements. Irradiation oversupply at the equator is the source for compensating air streams between the equator and the poles.

Besides the spatial compensation streams, less extensive air currents exist due to the influence of local areas of high and low pressure. The Coriolis force diverts the compensating streams between high and low pressure areas. Due to the rotation of the Earth, the air masses in the northern hemisphere are diverted to the right and in the southern hemisphere to the left. Finally, the air masses rotate around the low-pressure areas.

Wind resources are particularly high in *coastal areas* because wind can move unhindered across the smooth surface of the sea. Furthermore, temperature differences between water and land cause local compensating streams. The sunlight heats the land more quickly than the water during the day. The results are pressure differentials and compensating winds in the direction of the land. These winds can reach up to 50 km inland. During the night the land cools much faster than the sea; this causes compensating winds in the opposite direction.

Table 5.1 *Wind Speed Classification of the Beaufort Wind Scale*

Bf	v in m/s	Description	Effects
0	0–0.2	Calm	Smoke rises vertically
1	0.3–1.5	Light air	Smoke moves slightly and shows direction of wind
2	1.6–3.3	Light breeze	Wind can be felt. Leaves start to rustle
3	3.4–5.4	Gentle breeze	Small branches start to sway. Wind extends light flags
4	5.5–7.9	Moderate breeze	Larger branches sway. Loose dust on ground moves
5	8.0–10.7	Fresh breeze	Small trees sway
6	10.8–13.8	Strong breeze	Trees begin to bend, whistling in wires
7	13.9–17.1	Moderate gale	Large trees sway
8	17.2–20.7	Fresh gale	Twigs break from trees
9	20.8–24.4	Strong gale	Branches break from trees, minor damage to buildings
10	24.5–28.4	Full gale/storm	Trees are uprooted
11	28.5–32.6	Violent storm	Widespread damage
12	≥ 32.7	Hurricane	Structural damage

Note: Bf, Beaufort force; v, wind speed in m/s (1 m/s = 3.6 km/h = 2.24 mph)

Definition of wind force

In meteorology the *Beaufort scale* is often applied to give the wind force. This scale allows an approximate estimation of wind speed without complicated measurement systems; however, this scale is less useful for technical purposes. Therefore, the wind speed v given in the SI unit m/s is used instead. Table 5.1 compares the Beaufort classes with the corresponding wind speed values.

Wind speed distributions

Wind speed distributions are commonly used to indicate the annual available wind energy. These distributions are estimated using measurements, wind maps or computer analysis. Tables or statistical functions can give the distribution.

Figure 5.1 shows the *relative distribution* $h(v)$ of wind speed v in Karlsruhe, which is located in south Germany near the Black Forest. This distribution indicates how often a certain wind speed occurs. It is immediately obvious that the wind energy potential in Karlsruhe is very low. The sum of frequencies with wind speeds lower than 4 m/s is more than 70 per cent. In other words, for practical wind generator use, wind speeds above 4 m/s exist for only 30 per cent of the time.

The wind speed measurement interval can cause uncertainties in the estimation of wind speed frequency distributions. If the average of the wind

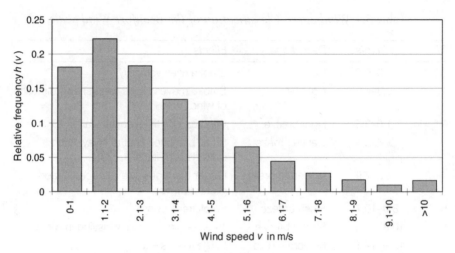

Figure 5.1 *Wind Speed Distribution for Karlsruhe in Inland Germany in 1991/1992*

speed is recorded only every 10 min or even every hour, further calculations on wind generator yield can produce high errors because the energy of the wind does not depend linearly on the wind speed. A solution that avoids this error is to record the average of the wind speed cubed.

The mean wind speed can be easily calculated:

$$\bar{v} = \sum h(v) \cdot v \tag{5.1}$$

However, the mean wind speed can only partly describe the potential of a site, because the wind distribution may be continuous wind or long calm periods interspersed with periods of very high wind speeds. The wind energy in these two cases can be totally different. Nevertheless, the mean wind speed is often used to give the site quality.

Good wind maps that show the mean wind speed exist for most countries (e.g. Troen and Petersen, 1989). In coastal areas, mean wind speeds of 6 m/s or more can be reached; in inland areas it can be below 3 m/s. Mountainous regions also offer good wind conditions. Today digital wind maps also exist and computer programs can estimate wind speeds even for locations where no measurements have been taken (e.g. Risø National Laboratory, 1987).

A wind speed frequency distribution gives much better information about the wind conditions of a certain site than the mean wind speed. The frequency distribution can be given as tables with wind speed intervals or as statistical functions. The most common statistical functions that are used for wind power calculations are the Weibull and the Rayleigh distributions.

The *Weibull distribution* of wind speed v with shape parameter k and scale parameter a is given by:

Table 5.2 *Weibull Parameters and Mean Wind Speed at a Height of 10 m for Various Locations in Germany*

Location	k	a	v in m/s	Location	k	a	v in m/s
Berlin	1.85	4.4	3.9	Munich	1.32	3.2	2.9
Hamburg	1.87	4.6	4.1	Nuremberg	1.36	2.9	2.7
Hannover	1.78	4.1	3.7	Saarbrücken	1.76	3.7	3.3
Helgoland	2.13	8.0	7.1	Stuttgart	1.23	2.6	2.4
Cologne	1.77	3.6	3.2	Wasserkuppe	1.98	6.8	6.0

Source: Christoffer and Ulbricht-Eissing, 1989

$$f_{\text{Weibull}}(v) = \frac{k}{a} \cdot \left(\frac{v}{a}\right)^{k-1} \cdot \exp\left(-\left(\frac{v}{a}\right)^{k}\right) \tag{5.2}$$

The shape and scale parameters depend on the site. Table 5.2 gives some example parameters for various German locations.

The mean wind speed can be estimated approximately from the Weibull parameters (Molly, 1990):

$$\bar{v} = a \cdot \left(0.568 + \frac{0.434}{k}\right)^{\frac{1}{k}} \tag{5.3}$$

The parameter a for $k = 2$ can be obtained from the mean wind speed:

$$a_{k=2} = \frac{\bar{v}}{0.886} \approx \frac{2}{\sqrt{\pi}} \cdot \bar{v} \tag{5.4}$$

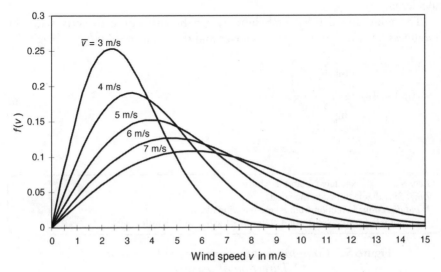

Figure 5.2 *Rayleigh Distributions for Different Mean Wind Speeds $\bar{v}$*

Substituting a in the Weibull distribution and using $k = 2$ results in the *Rayleigh distribution*:

$$f_{\text{Rayleigh}}(v) = \frac{\pi}{2} \cdot \frac{v}{\bar{v}^2} \cdot \exp\left(-\frac{\pi}{4} \cdot \frac{v^2}{\bar{v}^2}\right) \tag{5.5}$$

The Rayleigh distribution needs only the average wind speed as a parameter. Figure 5.2 shows Rayleigh distributions for different mean wind speeds.

Influence of surroundings and height

The wind speed is usually recorded at a height of 10 m. Changes in elevation can change the wind speed in a distance of only a few hundred meters. Hills or mountains influence the wind speed significantly. On top of a mountain or on the luff side, which is perpendicular to the wind (see Figure 5.3), the wind speed can rise to become double the uninfluenced value. In contrast, the wind speed is much lower on the lee side.

Obstacles, plants or hills near a wind generator site can slow the wind significantly. Single obstacles are no problem if the total rotor area is over three times higher than the obstacle or if there is sufficient distance between the wind generator and the obstacle. However, this distance can be up to 35 times the obstacle height. Without proper clearance, wind turbulence can reduce the usable wind energy.

The wind speed increases with the height from ground because the wind is slowed down by the roughness of the ground. Wind generators usually have hub heights of more than 10 metres. For the estimation of the wind potential, additional wind speed measurements at other heights are necessary. However, if the type of ground cover is known, the wind speed at other heights can be calculated.

The wind speed $v(h_2)$ at height h_2 can be calculated directly with the *roughness length*, z_0 of the ground cover and the wind speed $v(h_1)$ at height h_1:

$$v(h_2) = v(h_1) \cdot \frac{\ln\left(\dfrac{h_2 - d}{z_0}\right)}{\ln\left(\dfrac{h_1 - d}{z_0}\right)} \tag{5.6}$$

Lee: Side that is sheltered from the wind
Luff: Side that is turned to the wind
Wind direction: Direction from where the wind blows
90°: East wind, 180°: South wind, 270°: West wind, 360°: North wind

Figure 5.3 *Common Expressions for the Description of the Direction of the Wind*

Obstacles can cause a displacement of the boundary layer from the ground. This displacement can be considered by the parameter d. For widely scattered obstacles, parameter d is zero. In other cases d can be estimated as 70 per cent of the obstacle height.

The *roughness length* z_0 describes the height at which the wind is slowed to zero. In other words, surfaces with a large roughness length have a large effect on the wind. Table 5.3 shows the classification of various ground classes depending on the roughness length.

The following example shows the influence of the ground cover. The same wind speed $v(h_1) = 10$ m/s at a height $h_1 = 50$ m above different ground classes is assumed. Equation (5.6) is applied to calculate the wind speed $v(h_2)$ at a height of $h_2 = 10$ m. The displacement d for the boundary layer from the ground must be considered for higher obstacles in ground classes 6 to 8. Table 5.4 shows the calculated results. The wind speed decreases significantly with rising roughness lengths z_0; thus, it does not make any sense to install wind power plants in built-up areas or large forests. The wind speed also increases significantly with height. For instance, the wind speed at a height of 50 m is 30 per cent higher than at 10 m for ground class 4. This must be considered for the installation of large wind turbines. The usable wind speed at the top of large wind towers is much higher than at the common measurement height of 10 m. Wind turbines of the megawatt class come with hub heights of between 50 and 70 m for coastal areas (ground class 1 to 3) and even higher for inland areas with higher roughness lengths.

This example should not give the impression that the wind speed is already independent of the ground at a height of 50 m. The wind speed usually becomes independent of the height, where the wind becomes known as *geostrophic wind*, at heights significantly exceeding 100 m from the ground.

Finally, the power law of Hellmann is another relation for the vertical distribution of wind speeds.

Table 5.3 *Roughness Lengths* z_0 *for Different Ground Classes*

Ground class	Roughness length z_0 in m	Description
1 – Sea	0.0002	Open sea
2 – Smooth	0.005	Mud flats
3 – Open	0.03	Open flat terrain, pasture
4 – Open to rough	0.1	Agricultural land with a low population
5 – Rough	0.25	Agricultural land with a high population
6 – Very rough	0.5	Park landscape with bushes and trees
7 – Closed	1	Regular obstacles (woods, village, suburb)
8 – Inner city	2	Centres of big cities with low and high buildings

Source: Christoffer and Ulbricht-Eissing, 1989

Table 5.4 *Example of the Decrease in Wind Speed v(h₂) at Height h₂ = 10 m*
as a Function of the Ground Class for v(h₁) = 10 m/s at h₁ = 50 m

Ground class	z_0	d	$v(h_2)$ at h_2 = 10 m	Ground class	z_0	d	$v(h_2)$ at h_2 = 10 m
1	0.0002 m	0 m	8.71 m/s	5	0.25 m	0 m	6.96 m/s
2	0.005 m	0 m	8.25 m/s	6	0.5 m	3 m	5.81 m/s
3	0.03 m	0 m	7.83 m/s	7	1 m	5 m	4.23 m/s
4	0.1 m	0 m	7.41 m/s	8	2 m	6 m	2.24 m/s

With $z = \sqrt{h_1 \cdot h_2}$ and $a = \dfrac{1}{\ln \dfrac{z}{z_0}}$ Equation (5.6) becomes:

$$\frac{v(h_2)}{v(h_1)} = \left(\frac{h_2}{h_1}\right)^a \tag{5.7}$$

For z = 10 m and z_0 = 0.01 m, the parameter a is about 1/7; this equation is then called a 1/7 power law. However, this power law is only valid if the displacement d of the boundary layer from the ground is equal to zero.

UTILIZATION OF WIND ENERGY

Power content of wind

The *kinetic energy E* carried by a wind with speed v is given by the general equation:

$$E = \tfrac{1}{2} \cdot m \cdot v^2 \tag{5.8}$$

The *power P* that the wind contains is calculated by differentiating the energy with respect to time. For a constant wind speed v the power is:

$$P = \dot{E} = \tfrac{1}{2} \cdot \dot{m} \cdot v^2 \tag{5.9}$$

The density ρ and volume V determine the *mass*:

$$m = \rho \cdot V \tag{5.10}$$

The derivative with respect to time results in the *air mass flow*:

$$\dot{m} = \rho \cdot \dot{V} = \rho \cdot A \cdot \dot{s} = \rho \cdot A \cdot v \tag{5.11}$$

This mass of air with density ρ flows through an area A with speed v. Hence the power of the wind becomes:

Table 5.5 *Density of Air as a Function of the Temperature,*
$p = 1 \ bar = 1 \ hPa$

Temperature ϑ in °C	−20	−10	0	10	20	30	40
Density ρ in kg/m³	1.377	1.324	1.275	1.230	1.188	1.149	1.112

Source: data from VDI, 1993

$$P = \tfrac{1}{2} \cdot \rho \cdot A \cdot v^3 \tag{5.12}$$

The density of air ρ varies with the air pressure p and temperature ϑ. The density changes proportionally to the air pressure at constant temperature. Table 5.5 shows the density change for different temperatures at constant pressure.

Wind during a violent storm of Beaufort force 11 and a speed of 30 m/s at a temperature of 10°C reaches a power of 16.6 kW/m². With these high power densities, the devastation caused by violent storms is no surprise. However, the power of wind with a speed of 1 m/s is less than 1 W/m². Therefore, high mean wind speeds are essential for a good yield from a wind generator.

For the *utilization of wind power* a technical system such as a wind turbine should take as much power from the wind as possible. This turbine slows the wind from speed v_1 to speed v_2 and uses the corresponding power difference. If this happened in a pipe with rigid walls at constant pressure, the wind speed v_2 would change with the initial wind speed v_1, because the same amount of air that enters the pipe must leave it. Hence, the mass flow of the air before and after the wind turbine is the same.

Wind turbines slow down the wind when converting wind energy into electricity; however, the mass flow before and after the wind turbine remains constant. The wind flows through a larger cross-section after passing through the wind turbine as shown in Figure 5.4. For constant pressure and density ρ of the air, the mass flow is:

$$\dot{m} = \rho \cdot \dot{V} = \rho \cdot A_1 \cdot v_1 = \rho \cdot A \cdot v = \rho \cdot A_2 \cdot v_2 = const. \tag{5.13}$$

The wind speed

$$v = \tfrac{1}{2} \cdot (v_1 + v_2) \tag{5.14}$$

at the height of the wind turbine is the average of the wind speeds v_1 and v_2. The power P_T taken from the wind can be calculated from the difference in wind speeds:

$$P_T = \tfrac{1}{2} \cdot \dot{m} \cdot (v_1^2 - v_2^2) \ . \tag{5.15}$$

With $\dot{m} = \rho \bullet A \bullet v = \rho \bullet A \bullet \tfrac{1}{2} \bullet (v_1 + v_2)$, the expression becomes:

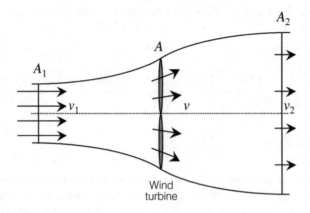

Figure 5.4 *Idealized Change of Wind Speed at a Wind Turbine*

$$P_T = \tfrac{1}{4} \cdot \rho \cdot A \cdot (v_1 + v_2) \cdot (v_1^2 - v_2^2) \qquad (5.16)$$

The power P_0 of the wind through the area A without the influence of the wind turbine is:

$$P_0 = \tfrac{1}{2} \cdot \rho \cdot A \cdot v_1^3 . \qquad (5.17)$$

The ratio of the power used by the turbine P_T to the power content P_0 of the wind is called the *power coefficient* c_P and is given by:

$$c_P = \frac{P_T}{P_0} = \frac{(v_1 + v_2) \cdot (v_1^2 - v_2^2)}{2 \cdot v_1^3} = \frac{1}{2} \cdot (1 + \frac{v_2}{v_1}) \cdot (1 - \frac{v_2^2}{v_1^2}) \qquad (5.18)$$

Betz has calculated the maximum power coefficient possible, which is called the ideal or *Betz power coefficient* $c_{P,Betz}$ (Betz, 1926).

With $\zeta = \dfrac{v_2}{v_1}$ and $\dfrac{dc_P}{d\zeta} = \dfrac{d\left(\tfrac{1}{2} \cdot (1+\zeta) \cdot (1-\zeta^2)\right)}{d\zeta} = -\tfrac{3}{2} \cdot \zeta^2 - \zeta + \tfrac{1}{2} = 0$,

the ideal wind speed ratio is $\zeta_{id} = \dfrac{v_2}{v_1} = \dfrac{1}{3}$.

When substituting the speed ratio in the equation of the power coefficient, the maximum power coefficient becomes:

$$c_{P,Betz} = \frac{16}{27} \approx 0.593 . \qquad (5.19)$$

If a wind turbine slows down air with an initial wind speed v_1 to one third of v_1 ($v_2 = 1/3 \cdot v_1$), the theoretical maximum power can be taken, and this maximum is about 60 per cent of the power content of the wind.

Real wind generators do not reach this theoretical optimum; however, good systems have power coefficients c_p between 0.4 and 0.5. The ratio of the used power P_T of the turbine to the ideal usable power P_{id} defines the efficiency η for the power utilization of the wind:

$$\eta = \frac{P_T}{P_{id}} = \frac{P_T}{P_0 \cdot c_{P,Betz}} = \frac{P_T}{\frac{1}{2} \cdot \rho \cdot A \cdot v_1^3 \cdot c_{P,Betz}} = \frac{c_P}{c_{P,Betz}} \qquad (5.20)$$

Drag devices

If an object is set up perpendicularly to the wind, the wind exerts a force F_D on the object. The wind speed v, the effective object area A and the *drag coefficient* c_D, which depends on the object shape, define the drag force:

$$F_D = c_D \cdot \tfrac{1}{2} \cdot \rho \cdot A \cdot v^2 \qquad (5.21)$$

Figure 5.5 shows drag coefficients for various shapes. With $P_D = F_D \cdot v$, the power to counteract the force becomes:

$$P_D = c_D \cdot \tfrac{1}{2} \cdot \rho \cdot A \cdot v^3 \qquad (5.22)$$

If an object moves with speed u by the influence of the wind in the same direction as the wind, the drag force is:

$$F_D = c_D \cdot \tfrac{1}{2} \cdot \rho \cdot A \cdot (v-u)^2 \qquad (5.23)$$

and the power used is:

$$P_T = c_D \cdot \tfrac{1}{2} \cdot \rho \cdot A \cdot (v-u)^2 \cdot u \qquad (5.24)$$

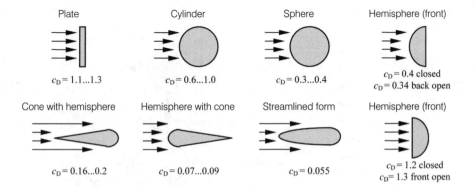

Source: Hering et al, 1992

Figure 5.5 *Drag Coefficients for Various Shapes*

The following example calculates approximately the used power of a cup anemometer that is used for the measurement of the wind speed v. It consists of two open hemispherical cups that rotate around a common axis. The wind impacts the front of the first cup and the back of the second cup (Figure 5.6).

The resulting force F consists of a driving and a decelerating component (Gasch and Twele, 2002):

$$F = c_{D1} \cdot \tfrac{1}{2} \cdot \rho \cdot A \cdot (v-u)^2 - c_{D2} \cdot \tfrac{1}{2} \cdot \rho \cdot A \cdot (v+u)^2 \qquad (5.25)$$

The used power is:

$$P_T = \tfrac{1}{2} \cdot \rho \cdot A \cdot \left(c_{D1} \cdot (v-u)^2 - c_{D2} \cdot (v+u)^2\right) \cdot u \qquad (5.26)$$

The ratio of the circumferential speed u to the wind speed v is called the tip speed ratio λ:

$$\lambda = \frac{u}{v} \qquad (5.27)$$

The tip speed ratio of drag devices is always smaller than one. Using the tip speed ratio, the power is:

$$P_T = \tfrac{1}{2} \cdot \rho \cdot A \cdot v^3 \cdot \left(\lambda \cdot \left(c_{D1} \cdot (1-\lambda)^2 - c_{D2} \cdot (1+\lambda)^2\right)\right) \qquad (5.28)$$

Hence, the *power coefficient of the cup anemometer* becomes:

$$c_P = \frac{P_T}{P_0} = \frac{P_T}{\tfrac{1}{2} \cdot \rho \cdot A \cdot v^3} = \lambda \cdot \left(c_{D1} \cdot (1-\lambda)^2 - c_{D2} \cdot (1+\lambda)^2\right) \qquad (5.29)$$

The maximum value of the power coefficient of the cup anemometer is about 0.073. This is much below the ideal Betz power coefficient of 0.593. The cup anemometer reaches its maximum power coefficient at a tip speed ratio of about 0.16, when the wind speed v is about six times higher than the circumferential speed u.

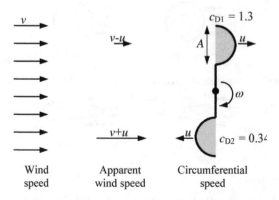

Wind   Apparent   Circumferential
speed   wind speed   speed

Figure 5.6 *Model of Cup Anemometer for the Calculation of Power*

The optimal power coefficient $c_{P,opt,D}$ of a drag device can be calculated using

$$c_P = \frac{P_T}{P_0} = \frac{\frac{1}{2} \cdot \rho \cdot A \cdot c_D \cdot (v-u)^2 \cdot u}{\frac{1}{2} \cdot \rho \cdot A \cdot v^3} = c_D \cdot (1 - \frac{u}{v})^2 \cdot \frac{u}{v} \tag{5.30}$$

Using $u/v = 1/3$ as well as the maximum drag coefficient of $c_{D,max} = 1.3$, this gives:

$$c_{P,opt,D} = 0.193 \tag{5.31}$$

This value is also much below the ideal value of 0.593. Therefore, modern wind turbines are lift devices, rather than drag devices, and these achieve much better power coefficients.

Lift devices

If wind, which circulates around a body, develops higher flow speeds along the upper surface than along the lower, an overpressure emerges at the upper surface and an underpressure at the lower. The result is a *buoyancy force*, according to Bernoulli:

$$F_L = c_L \cdot \frac{1}{2} \cdot \rho \cdot A_p \cdot v_A^2 \tag{5.32}$$

The buoyancy force is calculated using the *lift coefficient* c_L, the air density ρ, the apparent wind speed v_A and the projected body area A_p. Rotor blades of modern wind generators usually make use of the buoyancy force. The projected area

$$A_p = t \cdot r \tag{5.33}$$

of a rotor blade is defined by the chord t and span that is approximately equal to the rotor radius r.

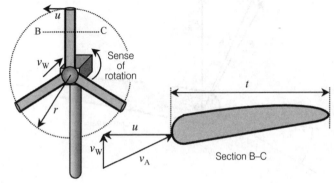

Figure 5.7 *Apparent Wind Speed* v_A *Resulting from the Real Wind Speed* v_W *and Rotor Motion*

Drag forces, which have been described in the section about drag devices (see p191), also have effects on lift devices:

$$F_D = c_D \cdot \tfrac{1}{2} \cdot \rho \cdot A_p \cdot v_A^2 \tag{5.34}$$

However, the buoyancy force on a drag device is much higher than the drag force. The ratio of both forces is called the *lift-drag ratio ε*:

$$\varepsilon = \frac{F_L}{F_D} = \frac{c_L}{c_D} \tag{5.35}$$

Some references also use the inverse ratio. Good rotor profiles can reach lift-drag ratios of up to 400.

The *apparent wind speed*:

$$v_A = \sqrt{v_W^2 + u^2} \tag{5.36}$$

used in the equations above is calculated from the real wind speed v_W and the circumferential speed u (see Figure 5.7). With the tip speed ratio $\lambda = u/v_w$, the apparent wind speed becomes:

$$v_A = v_W \cdot \sqrt{1 + \lambda^2} \tag{5.37}$$

Figure 5.8 shows the ratio of the drag force F_D to the buoyancy force F_L. Vector addition of both forces provides the resultant force:

$$F_R = F_D + F_L \tag{5.38}$$

The resultant force can be subdivided into an axial component F_{RA} and a tangential component F_{RT}. The tangential component F_{RT} of the resultant force causes the rotor to turn.

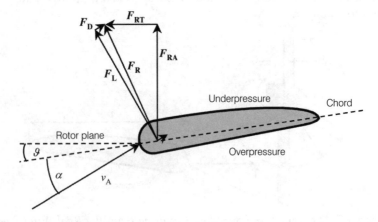

Figure 5.8 *Ratio of the Forces for a Lift Device*

Lift coefficient c_L and drag coefficient c_D vary significantly with the angle of attack α in °. For $\alpha < 10°$, the following approximation can be used (Gasch and Twele, 2002):

$$c_A \approx 5.5 \cdot \alpha \cdot \pi/180° \tag{5.39}$$

Pitching the rotor blade, i.e. changing the pitch angle ϑ that is shown in Figure 5.8, influences the angle of attack α and therefore the power coefficient c_P. The maximum of the power coefficient decreases significantly at high pitch angles and occurs at lower tip speed ratios. Pitch-controlled turbines use this effect: during starting of the wind turbine, high pitch angles are chosen. Pitching of the rotor blades can also limit the power at very high wind speeds. The drag coefficient c_D can be neglected at angles of attack lower than 15°. A lift device takes power P_T from the wind. With the power coefficient c_P and the power content P_0, the taken power is given by:

$$P_T = c_P \cdot P_0 = c_P \cdot \tfrac{1}{2} \cdot \rho \cdot A \cdot v_W^3 \tag{5.40}$$

The resultant force causes torque M. With

$$M = \frac{P_T}{\omega} = \frac{P_T \cdot r}{u} \tag{5.41}$$

the *torque* becomes:

$$M = c_P \cdot \frac{v_W}{u} \cdot \tfrac{1}{2} \cdot \rho \cdot A \cdot r \cdot v_W^2 \tag{5.42}$$

The torque M with the associated *moment coefficient*

$$c_M = c_P \cdot \frac{v_W}{u} = \frac{c_P}{\lambda} \tag{5.43}$$

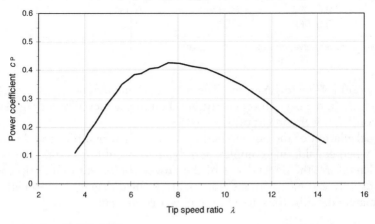

Source: data from Vestas, 1997

Figure 5.9 *Power Coefficient* c_P *as a Function of the Tip Speed Ratio* λ *for the Vestas V44-600-kW Wind Generator*

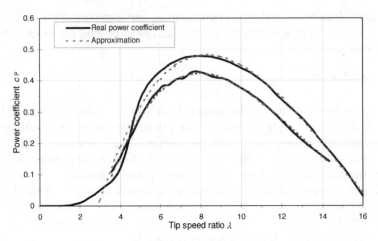

Figure 5.10 *Power Coefficients and Approximations using Third-degree Polynomials*

and $A = \pi\, r_2$, can be also described as follows:

$$M = c_M \cdot \tfrac{1}{2} \cdot \rho \cdot A \cdot r \cdot v_W^2 = c_M \cdot \tfrac{1}{2} \cdot \rho \cdot \pi \cdot r^3 \cdot v_W^2 \qquad (5.44)$$

If the torque M or power P of a wind generator as a function of the wind speed v_W is known, the power coefficient at constant speed can be calculated. Figure 5.9 shows the characteristics of the power coefficient and the tip speed

Table 5.6 *Parameters for the Description of the Power Coefficient Curves in Figure 5.10*

	a_3	a_2	a_1	a_0
Curve 1	0.00094	−0.0353	0.3841	−0.8714
Curve 2	0.00068	−0.0297	0.3531	−0.7905

Note: Curve 1 = lower curve; curve 2 = upper curve

ratio of a 600-kW wind generator. The maximum power coefficient of 0.427 is much closer to the Betz power coefficient than that achieved by a drag device (see section on power content of wind, p188).

The calculation of the power coefficient curve is very difficult and is only possible when considering complex aerodynamic conditions along the rotor blades. Therefore, the dependence of the power coefficient on the tip speed ratio is usually estimated by measurement. A third-degree polynomial can approximately describe the curve of the power coefficient:

$$c_p = a_3 \cdot \lambda^3 + a_2 \cdot \lambda^2 + a_1 \cdot \lambda + a_0 \qquad (5.45)$$

The coefficients a_3 to a_0 can be estimated with programs such as Matlab or MS-Excel from measurements. Figure 5.10 shows two real power coefficient curves and their *approximation by third-degree polynomials*. Table 5.6 shows the parameters of both curves.

WIND TURBINE DESIGN

The previous section explained in general how drag and lift devices can utilize wind power. This section describes technical solutions for this utilization. In the past, wind energy was mainly converted to mechanical energy; some modern wind pumping systems also use the mechanical energy directly.

However, today the generation of electricity is in higher demand; therefore, a wind rotor drives an electrical generator. Different concepts exist for the rotor design and are explained in the following sections.

Wind turbines with vertical rotor axis

Wind wheels and windmills with vertical axes are the oldest systems to exploit the wind. For more than 1000 years drag devices with vertical axes have been constructed. Today there are some modern wind generator concepts that also have vertical axes as shown in Figure 5.11.

Rotor concepts with vertical axes are:

- the Savonius rotor
- the Darrieus rotor and
- the H rotor.

The *Savonius rotor* works similarly to the above-described cup anemometer, using the drag principle. It has two semi-cylindrical blades that are open on

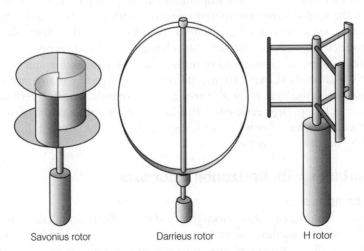

Savonius rotor   Darrieus rotor   H rotor

Figure 5.11 *Rotors with Vertical Axes*

opposite sides. Near the axis, the blades overlap so that the redirected wind can flow from one blade to the other. The rotor blades also utilize the lift principle so that the efficiency of the Savonius rotor is little better than that of simple drag devices. However, the efficiency is much worse than that of good lift devices, reaching maximum power coefficients of the order of 0.25 (Hau, 2000). Savonius rotors have the advantage that they can start at very low wind speeds. They are therefore used for ventilation purposes on buildings or utility vehicles. Besides poor efficiency, Savonius rotors have the disadvantage of a high material demand. Thus, they are not used in large systems.

The *Darrieus rotor* was developed by the Frenchman Georges Darrieus in 1929. The Darrieus rotor consists of two or three rotor blades that have the shape of a parabola. The profile of the rotor blades corresponds to lift devices so that the Darrieus rotor utilizes the lift principle. Due to its vertical axis the angle of attack at the Savonius rotor changes continuously. The efficiency of the Darrieus rotor is much above the efficiency of the Savonius rotor; however, it reaches efficiencies of only about 75 per cent of modern rotors with horizontal axes. A grave disadvantage of the Darrieus rotor is that is cannot start on its own: it always needs an auxiliary starting system that can be a drive motor or a coupled Savonius rotor.

A further development of the Darrieus rotor is the *H rotor* or H-Darrieus rotor. This rotor is also called the Heidelberg rotor after the company Heidelberg Motor. A permanent-magnet generator is directly integrated into the rotor structure and needs no gearbox. The rotor works in the same way as the Darrieus rotor as a lift device. The three rotor blades of the H rotor are attached vertically. Supports to the vertical axis help the rotor maintain its shape. The very robust H rotor was designed for the extreme weather conditions existing in high mountains or in Antarctica.

Wind power plants with vertical axes have some advantages. Their structure and their assembly are relatively simple. The electric generator and the gear as well as all electronic components can be placed on the ground. This simplifies the maintenance compared to rotors with horizontal axes. Rotors with vertical axes need not be oriented into the wind; therefore, they are perfectly suited for regions with very fast changes of wind direction.

However, these advantages have not resulted in a breakthrough for wind generators with vertical axes. Today, almost all wind power plants use rotors with horizontal axes; systems with vertical axes are only used for very special applications. The poorer efficiency and higher material demand of systems with vertical axes have been the deciding factors for the market dominance achieved by horizontal axis turbines to date.

Wind turbines with horizontal rotor axis

System components

Most wind turbines generating electricity today are horizontal axis machines. It is predominantly medium-sized enterprises that have pushed wind market developments. Wind power plants have reached a high technical level and

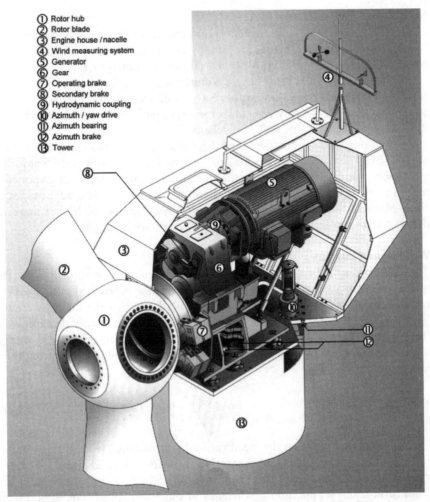

① Rotor hub
② Rotor blade
③ Engine house / nacelle
④ Wind measuring system
⑤ Generator
⑥ Gear
⑦ Operating brake
⑧ Secondary brake
⑨ Hydrodynamic coupling
⑩ Azimuth / yaw drive
⑪ Azimuth bearing
⑫ Azimuth brake
⑬ Tower

Figure 5.12 *Section through the Stall-controlled TW600 Wind Generator (600 kW Change-pole Asynchronous Generator, 43 m Rotor Diameter, 50–70 m Hub Height)*

current systems reach powers up to several megawatts, whereas the wind generators of the 1980s were in the power range below 100 kW.

A horizontal axis wind turbine generally consists of the following components (see Figure 5.12):

- rotor blades, rotor hub, rotor brake and if need be, a pitch mechanism
- electrical generator and if need be, a gearbox
- wind measurement system and yaw drive (azimuth tracking)
- nacelle, tower and foundation
- control, substation and mains connection.

Number of rotor blades

Modern horizontal axis wind generators can have one, two or three rotor blades. More than three blades are usually not used. The lower the number of rotor blades, the less material is needed during manufacturing.

Single-bladed rotors must have a counterweight on the opposite side of the rotor. Single-bladed rotors do not have a smooth motion and therefore exhibit a high material stress. Only very few prototypes with one rotor blade exist and it is not expected that this will change soon.

The optimal power coefficient of three-bladed rotors is slightly above that of two-bladed rotors. Three-bladed rotors have an optically smoother operation and hence visually integrate better into the landscape. The mechanical strain is also lower for three-bladed rotors. The advantages of three-bladed rotors compensate for the disadvantage of the higher material demand so that today mainly *three-bladed rotors* are built.

Wind speed ranges

The design tip speed ratio depends closely on the number of rotor blades. The maximum power coefficient of a three-bladed rotor is reached at a tip speed ratio of between 7 and 8, whereas for two-bladed rotors the figure is 10 and for a one-bladed rotor about 15. However, these values can vary depending on the system design. The optimum tip speed ratio λ_{opt} also defines the *design wind speed* along with the rotor radius r and the rotational speed of the rotor n in $\min^{-1}$.

$$v_D = \frac{u}{\lambda_{opt}} = \frac{2 \cdot \pi \cdot r \cdot n}{\lambda_{opt}} \tag{5.46}$$

For example, the optimum tip speed ratio of a three-blade wind turbine with a rotor radius of $r = 22$ m and a rotational speed of $n = 28$ $\min^{-1} = 0.467$ s^{-1} is $\lambda_{opt} = 7.5$, and therefore the design wind speed is $v_D = 8.6$ m/s. The wind turbine has its maximum efficiency at this wind speed. The design wind speed is very important for wind turbines with constant speed. Systems with variable speed can obtain the optimum efficiency at other wind speeds also. In this case, the design wind speed *range* should replace the design wind speed.

At very low wind speeds the operation of the wind turbine makes little sense. No power or only a very little power can be taken from the wind and the wind generator can even become a power consumer. Therefore, the rotor brake should stop the wind turbine below a predefined starting wind speed, or *cut-in wind speed* v_{cut-in}.

The design wind speed v_D was explained above; the rated or *nominal wind speed* v_N of a wind turbine is usually different. At the nominal wind speed, the wind turbine generates the rated power. The nominal wind speed is usually higher than the design wind speed. Above the nominal wind speed, the power of the wind turbine must be limited. If the wind speed becomes too high, the wind power plant can be overloaded and damaged. Therefore, wind turbines cut out at high wind speeds, $v_{cut-out}$: the rotor brakes stop the wind turbine

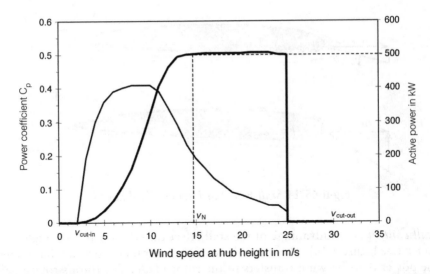

Source: data from Enercon, 1997

Figure 5.13 *Generator Active Power and Power Coefficient against Wind Speed for the 500-kW Enercon E-40 Wind Generator*

and the rotor is turned out of the wind if possible. The different wind speeds have the following typical ranges:

- cut-in wind speed v_{cut-in} = 2.5–4.5 m/s
- design wind speed v_D = 6–10 m/s
- nominal wind speed v_N = 10–16 m/s
- cut-out wind speed $v_{cut-out}$ = 20–30 m/s
- survival wind speed v_{life} = 50–70 m/s.

It must be considered that some companies give their generator curves for wind speeds at a height of 10 m, others at hub height (40–100 m). Figure 5.13 shows the generator active power and the power coefficient for a variable-speed 500-kW wind turbine as a function of the wind speed.

Limiting power output and storm protection

The power that can be taken from the wind varies with the wind speed. After reaching the nominal power, the power should remain constant for wind speeds greater than the nominal wind speed because the turbine and generator cannot handle more power. Therefore, a wind power plant must limit the power with one of the two following methods:

- stall control
- pitch control.

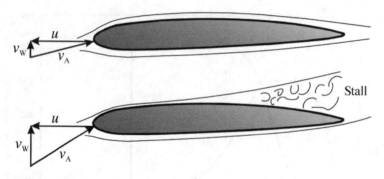

Figure 5.14 *Stall Effect at Higher Wind Speeds*

Stall control takes advantage of the stall effect that occurs at high angles of attack (see Figure 5.14). This effect destroys the buoyancy force and thus limits the power that the wind transfers to the rotor blade. The rotor's rotational speed n and the circumferential speed u remain constant for stall-controlled wind generators. Stall control is achieved by increasing the angle of attack at higher wind speeds v_W. The rotor blades do not pitch, i.e. the pitch angle remains constant; stall control can be realized by construction measures without advanced technical requirements. The disadvantage of stall control is the low possibility to influence operation because stall control is purely passive. The maximum power of a newly designed rotor blade is not easy to estimate because a mathematical description of the stall effect is rather difficult. After reaching the maximum power, the power output of stall-controlled wind turbines decreases again and does not remain on a constant level as shown in Figure 5.13.

Many manufacturers of wind turbines prefer *pitch control*, although the technical effort is much higher than for stall control. However, since pitch control is an active control, it can be adjusted to suit the conditions, in contrast to stall-controlled systems. Pitch control directly increases or lowers the pitch angle of the rotor and therefore the angle of attack. The rotor blade is turned into the wind at higher wind speeds (see Figure 5.15), lowering the angle of attack and actively decreasing the power input of the rotor blade. Pitch-controlled wind turbines are more difficult to manufacture because the rotor blades must twist inside the rotor hub. Small systems often use mechanically controlled pitch mechanisms using centrifugal forces. An electric motor moves the rotor blade to the desired position in large systems.

If the wind generator is stopped due to storm protection, the pitch control can pitch the rotor blade towards the feather position. This reduces the power input and avoids damage to the wind turbine. Stall-controlled systems often have additional aerodynamic brakes. For instance, the rotor tip can bend. During storms, the tip bends by 90° and slows the wind turbine.

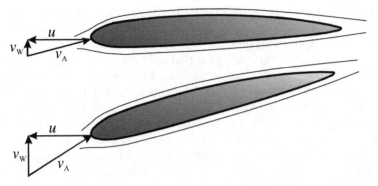

Figure 5.15 *Rotor Blade Positions for Different Wind Speeds for a Pitch-controlled System*

Yawing

Horizontal axis wind turbines must always follow the direction of the wind, in contrast to vertical axis wind turbines. The orientation of the rotor blades must be chosen so that the rotor blades face the wind at the optimal angle. This can be a problem for pitch-controlled wind turbines if the direction of the wind changes very fast or is gusty. Hence, high power fluctuations can occur and must be cushioned by changes in the rotor speed.

The position of the rotor can be upwind or downwind. The position of the rotor relative to the tower for *upwind turbines* is before the tower in the direction of the wind, and for *downwind turbines* it is behind. The disadvantage of the downwind rotor is that the rotor blades have to continually pass the sheltered zone of the tower. This produces high mechanical strains and noise emissions due to turbulence from the tower and the nacelle. Therefore, most large wind turbines are upwind turbines. Downwind turbines have the advantage that wind pressure adjusts the rotating rotor blades optimally to the wind. Small wind turbines can use wind vanes for passive yawing. The wind vane moves the rotor of an upwind turbine always to a position perpendicular to the wind.

To move a horizontal axis wind turbine in the yaw direction, the whole nacelle with rotor, gearbox and generator must be movable on top of the tower. Wind measuring equipment on the nacelle estimates the wind speed and direction and a control unit decides when an electric or hydraulic *yaw drive* moves the nacelle and rotor azimuth. When the nacelle reaches its optimal position, azimuth brakes hold this position. In reality there are always small deviations of the direction of the wind and the optimal position of the rotor. This deviation is called the yaw angle and is usually about 5°.

Tower, Foundations, Gearbox, Nacelle and Generator

The *tower* is one of the most important parts of the wind power plant. It must hold the nacelle and the rotor blades. Higher tower heights can increase the wind generator output because the wind speed rises with height. Early in their history, wind towers were lattice towers. The advantage of lattice towers is the lower material demand and costs. Because lattice towers are now considered to be visually unappealing, they are only rarely used today. Very small wind generators use simple towers with guys that hold the tower. Larger wind power plants use tubular towers, which are made of steel or concrete with circular cross-section. Modern towers can reach heights up to 100 m and even more. A 60-m-high steel tower with a diameter of 3.65 m at the bottom and 2 m at the top has a weight of 55 t. Hence, transport and final assembly of tower and rotor blades can become rather difficult for large systems, especially in rough terrain. Carefully designed foundations must hold these high masses so as to withstand even severe storms. There are well-documented cases of badly-designed wind power plants that have fallen during storms.

The nacelle of the wind power plants carries the rotor bearings, the *gearbox* and the electric generator. Since most generator designs need high rotational speed, a gearbox must adapt the rotor speed to the generator speed. However, a gearbox has a lot of disadvantages: it causes higher costs, friction losses, noise pollution and higher maintenance efforts. Gearless wind power plant concepts with specially designed generators try to avoid these disadvantages. These generators must have a high number of magnetic poles so that the generator can also produce electricity at low rotor speeds. However, a high number of poles enlarge the generator significantly and therefore increase its cost.

ELECTRICAL MACHINES

Electrical generators for converting mechanical energy produced by the wind to electricity are the heart of a wind power station. Therefore, this section is dedicated to electrical machines. Before their application in wind power can be described, some basics must be explained.

The type of current is one important characteristic feature of electrical machines. Major current types are:

- direct current (DC)
- alternating current (AC)
- impulse current
- three-phase current.

DC machines are the oldest machines, because their principle is easy to understand. Today they are used for many applications, for instance windscreen wiper motors in cars. The universal motor that works with AC is mainly used for household appliances. Stepping motors work with pulsed current, for instance in computer printers.

However, *three-phase machines* are the best suited for electricity generation at wind power plants and therefore only these are described here. Three-phase machines can be subdivided into *asynchronous machines* and *synchronous machines*.

Alternating current, AC

Three-phase machines work with a three-phase AC. Before three-phase machines can be explained in detail, a short introduction into AC calculations is given here.

An alternating voltage dependent on time is defined as:

$$v(t) = \hat{v} \cdot \cos(\omega \cdot t + \varphi_v) \tag{5.47}$$

with amplitude $\hat{v}$, zero phase angle φ_v of the voltage and the angular frequency ω given by:

$$\omega = 2\pi \cdot f \tag{5.48}$$

The associated *current* i with the zero phase angle φ_i and the amplitude $\hat{i}$ is:

$$i(t) = \hat{i} \cdot \cos(\omega \cdot t + \varphi_i) \tag{5.49}$$

The phase angle φ between current and voltage is:

$$\varphi = \varphi_v - \varphi_i \tag{5.50}$$

If the phase angle is positive, the voltage is ahead of the current (leading). If the phase angle is negative, the voltage is behind (lagging). Figure 5.16 shows the curve of current and voltage as a function of time. With zero phase angles $\varphi_v = 0$ and $\varphi_i = -\pi/4$, the phase angle of this example is $\varphi = +\pi/4$, i.e. the voltage is ahead of the current.

Another common description is the vector diagram of the amplitudes. Here, the amplitude of the voltage is usually the vertical reference vector. The other amplitudes are added relatively with their phase angle. The amplitudes of the current and voltage of the example above are shown in the vector diagram of the amplitudes in Figure 5.16.

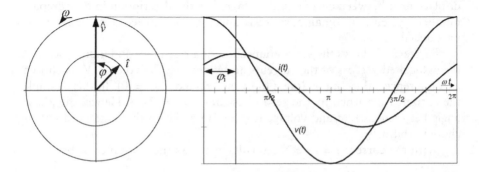

Figure 5.16 *Current and Voltage as a Function of Time and Vector Diagram of the Amplitudes $\hat{i}$ and $\hat{v}$ ($\varphi = \pi/4$)*

For a description of alternating quantities, the definition of an average value is interesting. Since the positive and negative parts of a sinusoidal oscillation cancel each other out when calculating the arithmetic average value, the *root mean square value* (*rms*) is used in electrical engineering. The definition of the root mean square value of a function $v(t)$ with the periodic time $T = 1/f$ is as follows:

$$v_{\text{rms}} = V = \sqrt{\frac{1}{T} \cdot \int_0^T v^2(t)\, dt} \tag{5.51}$$

For sinusoidal currents and voltages the rms values are:

$$V = v_{\text{eff}} = \frac{\hat{v}}{\sqrt{2}} \approx 0.707 \cdot \hat{v} \tag{5.52}$$

$$I = i_{\text{eff}} = \frac{\hat{i}}{\sqrt{2}} \approx 0.707 \cdot \hat{i} \tag{5.53}$$

The vector diagram of the rms values is similar to that of the amplitudes shown in Figure 5.16, except that the lengths of the vectors are shorter. The vector diagram is also used in mathematics for complex numbers. Therefore, the vectors of current and voltage can be interpreted as complex quantities. However, the real axis (Re) of the vector diagram for the currents and voltages is drawn vertically, whereas in mathematics it is usually drawn horizontally. *Complex quantities* are underlined in the following expressions.

With the complex voltage $\underline{V} = V \cdot e^{j\varphi_v} = V \cdot e^{j0} = V$ of the example above, the complex value of the rms value of the current with the phase angle φ_i and the imaginary unit j ($j^2 = -1$) becomes:

$$\underline{I} = I \cdot e^{j\varphi_i} \tag{5.54}$$

Electronic components such as capacitors or inductors cause a phase displacement between current and voltage. For the description in the complex number system, an imaginary resistance, the so-called reactance X, is introduced.

Figure 5.17 shows the series connection of a resistance and inductance and the associated vectors of the currents and voltages. The voltage $\underline{V}_1$ is chosen as a reference value and is drawn onto the real axis ($\varphi_v = 0$). In this example the current I is turned by the zero phase angle $\varphi_i = 3\pi/4$. Hence, the phase angle between current and voltage is $\varphi = -3\pi/4$. This value in the example is chosen arbitrarily.

With the current $\underline{I} = I \cdot e^{j\varphi_i}$, the voltage across the resistance R becomes:

$$\underline{V}_R = R \cdot \underline{I} = R \cdot I \cdot e^{j\varphi_i} \tag{5.55}$$

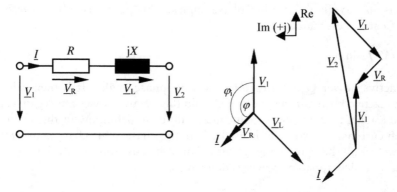

Figure 5.17 *Series Connection of Resistance and Inductance with Vector Diagram*

The vector of the voltage $\underline{V}_R$ has the same direction as the vector of the current $\underline{I}$. The voltage across an inductance L with reactance $X = \omega L$ and $j = e^{j\frac{\pi}{2}}$, becomes:

$$\underline{V}_L = j\omega L \cdot \underline{I} = jX \cdot \underline{I} = e^{j\frac{\pi}{2}} \cdot X \cdot I \cdot e^{j\varphi_i} = X \cdot I \cdot e^{j(\frac{\pi}{2}+\varphi_i)} \qquad (5.56)$$

The voltage V_2 can now be calculated as:

$$\underline{V}_2 = \underline{V}_1 - \underline{V}_L - \underline{V}_R \qquad (5.57)$$

The translation of the voltage vectors $\underline{V}_R$ and $\underline{V}_L$ provides the vector $\underline{V}_2$ that closes the loop in the vector diagram.

The instantaneous power $p(t)$ is calculated similarly to the DC power as:

$$p(t) = v(t) \cdot i(t) \qquad (5.58)$$

The *active power* P is the difference between the positive and negative areas defined by the curve $p(t)$ and the horizontal time axis. Using

$$P = \frac{1}{T} \int_0^T v(t) \cdot i(t) \cdot dt, \qquad (5.59)$$

the active power P of a harmonic voltage and current curve with the phase angle φ becomes:

$$P = \frac{1}{2} \cdot \hat{v} \cdot \hat{i} \cdot \cos\varphi = V \cdot I \cdot \cos\varphi \qquad (5.60)$$

If the phase displacement between current and voltage is $\pm\pi/2$ ($\pm 90°$), the positive areas are equal to the negative areas. Thus, the active power becomes zero. However, this does not mean that there is really no power. The remaining power oscillates between consumer and generator. A lower level of oscillating power is also available at other phase angles. This power is called *reactive power Q* and is estimated by:

$$Q = V \cdot I \cdot \sin \varphi \tag{5.61}$$

The reactive power Q is positive for positive phase angles. It is then called lagging reactive power. If the phase angle and the reactive power are negative, it is called leading reactive power. Sometimes a different definition for the relation between currents and voltages is used where the phase angle changes by 180° and therefore all the signs change as well. This can cause some confusion.

The so-called apparent power is defined as:

$$\underline{S} = P + j \cdot Q \quad \text{and} \quad |\underline{S}| = S = \sqrt{P^2 + Q^2} = V \cdot I. \tag{5.62}$$

The unit of the active power P is W (Watts). The unit of the reactive and apparent power is also the product of the units of voltage and current. To avoid confusion and to distinguish between the three power types, other units are defined for the reactive and apparent power. The reactive power Q is expressed by the unit var (Volt Ampere reactive) and the apparent power S with the unit VA (Volt Ampere).

The *power factor*

$$\cos \varphi = \frac{P}{S} \tag{5.63}$$

describes the ratio of the active power P to the apparent power S. Since the cosine of negative and positive phase angles provides the same value, the power factor is annotated with the supplemental labels 'inductive' or 'capacitive' in many cases. This indicates clearly if the phase angle is positive or negative.

Rotating field

If an electric current flows through a wire, it causes a magnetic field H as shown in Figure 5.18 for a wire and a coil.

The magnetic field strength H at a distance r from an active wire with electric current I is:

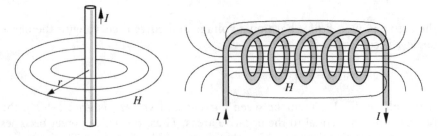

Figure 5.18 *Magnetic Fields* H *Produced by an Electric Current in a Wire and Coil*

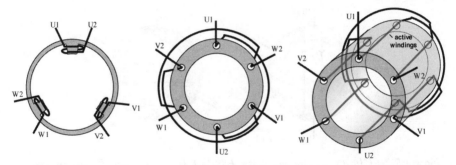

Figure 5.19 *Left: Cross-section through a Stator with Three Coils Staggered by 120° for the Generation of a Rotating Field (Concentrated Winding); Middle: Cross-section; Right: Three-dimensional Drawing of an Integrated Rotating Field Winding (Distributed Winding)*

$$H = \frac{I}{2 \cdot \pi \cdot r} \tag{5.64}$$

Besides the magnetic field strength, the *magnetic induction B* can be estimated using the magnetic permeability of free space μ_0:

$$\mu_0 = 4 \cdot \pi \cdot 10^{-7} \frac{\text{Vs}}{\text{Am}} \approx 1.257 \cdot 10^{-6} \frac{\text{Vs}}{\text{Am}} \tag{5.65}$$

and the material-dependent permeability coefficient μ_r from the following:

$$B = \mu_0 \cdot \mu_r \cdot H \tag{5.66}$$

The magnetic field of active wires is used for generating a rotating field. Therefore, a three-phase AC flows through three coils that are staggered by 120°. Figure 5.19 shows three coils U, V and W that are staggered by 120° with the connections U1, U2, V1, V2, W1 and W2.

If three ACs, that are staggered temporally by 120°, are fed into 120° spatially staggered *three-phase windings*, a rotating field emerges. Figure 5.20 explains this for two points in time. The currents at stages I ($\omega_t = 0$) and II ($\alpha_t = \pi/2$) can be estimated from Figure 5.21. The currents create a magnetic field as explained above. The north pole of the magnetic field rotates clockwise by 90° between stages I and II. If the magnetic field is constructed in the same manner for other points in time, it can be realized that the magnetic field changes its direction continuously. This changing magnetic field is also called a *rotating field*.

A magnetic needle inside a stator with three-phase windings would follow the rotating field and therefore rotate continuously. The frequency of the current defines the rotational speed. Hence, the magnetic needle would make one rotation during every period of the current. At the European mains

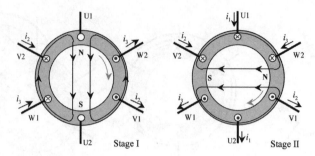

Figure 5.20 *Change in the Magnetic Field at Two Different Points in Time (Stage I and Stage II) when Supplying Three Sinusoidal Currents that are Temporally Staggered by 120°*

frequency of $f = 50$ Hz, the synchronous speed n_S, i.e. the speed of the rotating magnetic field, is $n_S = 50$ s^{-1} = 3000 min^{-1}. For a frequency p of 60 Hz, n_s becomes 3600 min^{-1}.

The magnetic field of the stator with three-phase windings in Figure 5.20 has only two poles N and S, i.e. it has one pole pair ($p = 1$). Stators and windings can also be produced with more pole pairs. When doubling the pole pairs the rotational speed halves if the mains frequency remains constant. The mains frequency f and the pole pair number p define the *synchronous speed*:

$$n_S = \frac{f}{p} \tag{5.67}$$

The pole pitch can be calculated with the pole pair number p and the diameter d of the stator:

$$\tau_p = \frac{d \cdot \pi}{2 \cdot p} \tag{5.68}$$

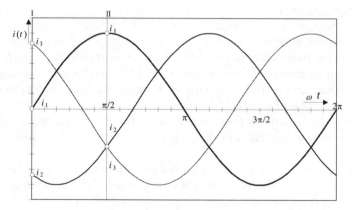

Figure 5.21 Three-phase Currents to Generate a Rotating Field

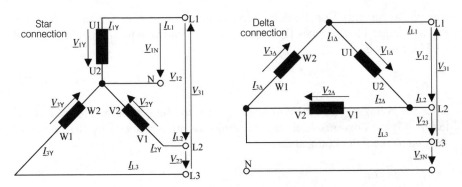

Figure 5.22 *Principle of Star and Delta Connections*

For a distance x along the circumference, the temporal and spatial staggered magnetic induction B with the amplitude $\hat{B}$ becomes:

$$B(x,t) = \hat{B} \cdot \sin(\frac{\pi \cdot x}{\tau_p} - \omega \cdot t) \qquad (5.69)$$

The six connections U1, U2, V1, V2, W1 and W2 of the three windings are connected either to a star or to a delta connection as shown in Figure 5.22. This reduces the number of external connections to the three phases L1, L2 and L3. These three phases are often also called R, S and T. The neutral conductor N can be added as reference; however, a neutral or zero conductor is not necessary for three-phase machines.

The rms values of the phase-to-phase voltages and the phase to neutral conductor voltages are related as follows:

$$V = V_{12} = V_{23} = V_{31} = \sqrt{3}\, V_{1N} = \sqrt{3}\, V_{2N} = \sqrt{3}\, V_{3N} \qquad (5.70)$$

In the European grid, the rms value of the voltage between phase and neutral conductors is 230 V and between two phases $\sqrt{3} \cdot 230$ V = 400 V. All voltages are phase-shifted by $2\pi/3$.

For the *star connection*, the voltages V_Y at the windings are equal to the corresponding phase-to-neutral conductor voltages. The voltages V_D at the windings of a *delta connection* are equal to the phase-to-phase voltages V. Hence, the connection between the rms voltages is:

$$V = V_\Delta = \sqrt{3} \cdot V_Y \qquad (5.71)$$

The currents I_Y in the windings of a star connection are the same as in the phases I:

$$I = I_L = I_Y \qquad (5.72)$$

The phase currents I of the delta connection are split between the currents in the windings. This reduces the rms currents I_Δ in the windings by a factor of $\sqrt{3}$ The result is:

$$I = \sqrt{3} \cdot I_\Delta \tag{5.73}$$

Since the voltages V_Y at the windings of the star connection are reduced by the factor $\sqrt{3}$, the currents I_Y in the windings are also reduced by the factor $\sqrt{3}$ relative to the delta connection:

$$I_\Delta = \sqrt{3} \cdot I_Y \tag{5.74}$$

Thus, the phase currents I of a star and a delta connection with the same windings and the same phase-to-phase voltages are not identical.

The total active power of a symmetrical three-phase system in a star connection is given by the sum of the active powers of all three windings:

$$P = 3 \cdot V_Y \cdot I_Y \cdot \cos\varphi = \sqrt{3} \cdot V \cdot I \cdot \cos\varphi \tag{5.75}$$

The total active power of a delta connection is calculated similarly:

$$P = 3 \cdot V_\Delta \cdot I_\Delta \cdot \cos\varphi = \sqrt{3} \cdot V \cdot I \cdot \cos\varphi \tag{5.76}$$

The measurement of voltage and current of only one phase is therefore sufficient to estimate the total power of a symmetrical connection. With $V_\Delta = \sqrt{3} \cdot V_Y$ and $I_\Delta = \sqrt{3} \cdot I_Y$, the correlation between the active power of a star and a delta connection becomes:

$$P_\Delta = 3 \cdot P_Y \tag{5.77}$$

In other words, the power input of a delta connection is three times higher than the associated star connection.

The *reactive power Q* and the *apparent power S* of delta and star connections are calculated similarly to the active power. Hence, they are:

$$Q = \sqrt{3} \cdot V \cdot I \cdot \sin\varphi \quad \text{and} \tag{5.78}$$

$$S = \sqrt{3} \cdot V \cdot I \tag{5.79}$$

Synchronous machines

Design of synchronous machines

A synchronous machine consists of a *stator* and a *rotor*. The stator is the stationary part of the machine with the three-phase windings to produce the rotating field as described in the previous section. The casing is made of

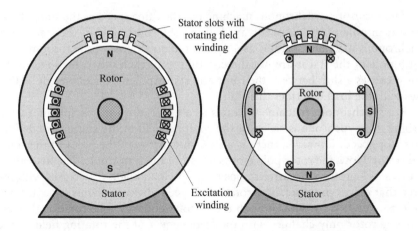

Figure 5.23 *Cross-section through a Synchronous Machine; Left: Cylindrical Rotor, Right: Salient-pole Rotor*

stacked sheets containing the three-phase windings of the stator, usually located inside open slots along the inner borehole (distributed winding).

As described before, the rotating field would cause a magnetic or compass needle inside the stator to rotate with the frequency of the rotating field. However, in the borehole of the stator of a synchronous generator, it is not a magnetic needle but a rotor that can be driven by the rotor blades of a wind turbine. Such a rotor must be magnetic so that it can follow the frequency of the rotating field. Permanent magnets or rotor windings with a DC current produce the magnetic field of a synchronous machine rotor. The rotor windings are also called *excitation windings*. The DC current that flows in these windings is fed from outside through slip rings.

There are two types of synchronous machine rotors, which are shown in Figure 5.23. They are cylindrical and salient-pole rotors. The *cylindrical rotor*, or turbo rotor, has a solid drum. It has slots in the longitudinal direction that contain the excitation windings. A cylindrical rotor can resist centrifugal forces better due to its massive construction. However, the material requirements of cylindrical rotors are also higher.

A *salient-pole rotor* has two or more salient poles. These rotors can have two poles, four poles (see Figure 5.23) or even more. The theoretical description of a salient-pole rotor is much more complicated than that of a cylindrical rotor since its construction causes asymmetries. This section describes cylindrical rotors only. For a description of salient-pole rotors and further details of electrical machines see the specialized literature (e.g. Hindmarsh, 1995; Fitzgerald et al, 2002).

The rotational speed $n_S = f_1 / p$ of the stator field depends on the frequency f_1 of the three-phase current and the pole pair number p of the stator as described before. The rotational speed for a frequency of 50 Hz and two poles ($p = 1$) is $n_S = 3000$ min^{-1}. For 60 Hz it becomes 3600 min^{-1}.

The rotor of a synchronous machine has the same speed as the stator field. The north pole of the rotor always follows the south pole of the stator. If a synchronous machine operates as a motor, the north pole of the rotor and the south pole of the stator are not directly on top of each other; the load of the motor causes a shift between the rotor and stator poles. The load angle ϑ that rises with the load describes this shift.

If a synchronous machine operates as a generator, for example in a wind turbine, the synchronous speed of the rotating field in the stator also defines the rotor speed. However, there is also a shift between the poles of the rotor and stator. Now, the rotor pole moves ahead of the stator pole. The load angle changes its sign from negative to positive. The load angle ϑ increases with the force that drives the rotor. The rotational speed always remains constant, i.e. the rotor runs synchronously with the stator frequency. The rotational speed n_s of the rotor only changes with the frequency f of the rotating field or the pole pair number p.

Electrical description of a synchronous machine

All three phases of the stator windings produce a main field that is connected to the rotor. The useful inductance L_h and the useful reactance $X_h = 2 \cdot \pi \cdot f_1 \cdot L_h$ reflect this. Besides the useful field there are leakage fields that are not connected to the rotor. The leakage reactance X_σ describes these leakages. The rotor field of a rotating rotor induces a voltage leak in the stator. This voltage is also called the *synchronous internal voltage* $\underline{V}_p$. The rms value of the induced voltage is proportional to the excitation current I_E in the rotor:

$$V_p \sim I_E \; : \tag{5.80}$$

in other words, the induced voltage can be adjusted by changing the excitation current.

Besides the voltage drop over the reactances there is an ohmic voltage drop at the stator resistance R_1. Hence, the equation for the equivalent circuit that describes the connection between the stator current $\underline{I}_1$ and stator voltage $\underline{V}_1$ for one phase becomes:

$$\underline{V}_1 = \underline{V}_p + \underline{I}_1 \cdot R_1 + \underline{I}_1 \cdot j(X_h + X_\sigma) \tag{5.81}$$

The influence of the stator resistance R_1 is low for large machines so that it can be neglected. The useful reactance X_h and the leakage reactance X_σ can be united as the synchronous reactance:

$$X_d = X_h + X_\sigma \tag{5.82}$$

This simplifies the equation of the one-phase *equivalent circuit* (Figure 5.24):

$$\underline{V}_1 = \underline{V}_p + \underline{I}_1 \cdot j X_d \tag{5.83}$$

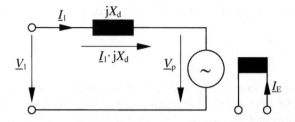

Figure 5.24 *Simple Equivalent Circuit (R$_1$ = 0) of a Cylindrical Rotor Machine for One Phase*

A synchronous machine can generate reactive power if desired. This is the most significant advantage of the synchronous machine compared to the asynchronous machine described later. The stator current of a synchronous machine can run through all phase angles. This is also called *four-quadrant operation.* Figure 5.25 shows the vector diagrams of a cylindrical rotor machine for four different operation conditions.

Depending on the phase angle, states are described as *underexcitation* and *overexcitation.* If a synchronous machine is underexcited, it behaves like a coil and absorbs reactive current. An overexcited synchronous machine has the same behaviour as a capacitor and generates lagging reactive current.

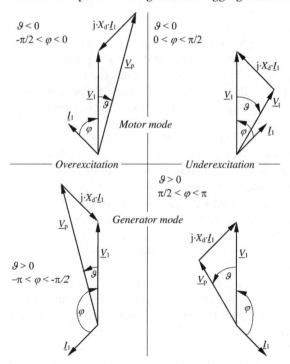

Figure 5.25 *Vector Diagrams of a Synchronous Machine with Cylindrical Rotor*

Angular relations in the vector diagram provide the following relation:

$$I_1 \cdot \cos \varphi = -\frac{V_p}{X_d} \cdot \sin \vartheta \qquad (5.84)$$

Hence, the phase current I_1 and the phase angle φ depend on load and excitation. The load angle ϑ increases with rising load, i.e. torque M. The synchronous internal voltage V_p depends only on the excitation current.

The electrical active power P_1 of the stator

$$P_1 = 3 \cdot V_1 \cdot I_1 \cdot \cos \varphi \qquad (5.85)$$

can be calculated from the active powers of the three winding phases of the stator. With the synchronous speed n_S of the rotating field and

$$M = \frac{P_1}{2 \cdot \pi \cdot n_S} \qquad (5.86)$$

the *torque* that is associated with the active power becomes:

$$M = -\frac{3 \cdot V_1}{2 \cdot \pi \cdot n_S} \cdot \frac{V_p}{X_d} \cdot \sin \vartheta \qquad (5.87)$$

According to this definition, the power P and torque M are negative in generator mode with negative load angle ϑ and the machine produces power. The rotor torque is higher than the torque that corresponds to the active power because friction, gearbox and other losses reduce the rotor torque.

Figure 5.26 shows the curve of the torque M over the load angle ϑ. A synchronous machine produces the theoretical maximum torque, called the pull-out torque, M_p, at load angles of $\pm\pi/2$. The magnitude of this torque changes with the internal voltage and therefore with the excitation current. If the load torque of the machine increases above the pull-out torque M_p, the machine falls out of step. If it operates in motor mode it stops. If it operates in generator mode, the rotor runs faster than the rotating field in the stator and the machine overspeeds, in which case high centrifugal forces could destroy the rotor of a wind generator. Reliable safety systems such as aerodynamic brakes must avoid this critical operation mode.

Synchronization

Before a synchronous generator can be connected to the mains it must be synchronized with the mains. Up to now it has been assumed that the stator frequency is equal to the mains frequency. During the start-up of a synchronous generator, the rotor speed and therefore the stator frequency is different from the mains frequency. If a synchronous generator were to be connected immediately to the mains, strong compensation reactions would appear such as very high compensation currents. This would cause

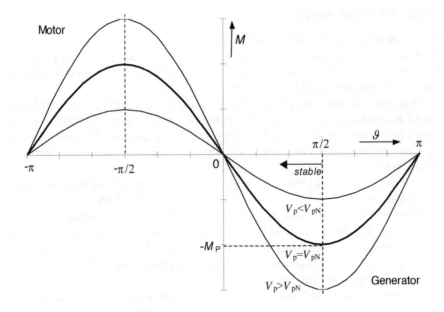

Figure 5.26 *Curve of the Torque of a Synchronous Machine with Cylindrical Rotor as a Function of the Load Angle ϑ and the Internal Voltage* V_p

unacceptable strain on the mains and could even damage the generator. Therefore, the following *synchronization conditions* must be fulfilled before connecting a synchronous generator to the mains:

- same phase sequence of mains and stator (same sense of rotation)
- equality of the frequencies of stator and mains
- equality of the voltage amplitude of stator and mains
- equality of phase characteristic of the stator and mains voltage.

The synchronization of a modern synchronous generator usually happens fully automatically. The following synchronization conditions must be fulfilled as a minimum (VDEW, 1994):

- voltage difference $\quad\quad\quad \Delta V \quad < \pm\, 10\%\ V_N$
- frequency difference $\quad\quad \Delta f \quad < \pm\, 0.5$ Hz
- phase angle difference $\quad\ \Delta\varphi \quad < \pm\, 10°$.

Asynchronous machines, which are described in the next section, do not need such a synchronization process.

Asynchronous machines

Design and operating conditions

The *stator* of an asynchronous machine, i.e. the stationary part, is in principle the same as a synchronous machine: three spatially distributed windings generate the rotating field in the stator.

However, the *rotor* of an asynchronous machine is totally different from that of a synchronous machine. The rotor of an asynchronous machine has neither permanent magnets nor DC windings. There are two different types of rotor for asynchronous machines: the squirrel-cage, or cage rotor, and the slipring rotor.

The three-phase windings of a *cage rotor* are made of bars that are inside slots in the rotor. Short-circuiting rings connect the ends of the bars.

The ends of the windings of a *slipring rotor* are only connected internally on one side. The beginnings of the windings are connected via sliprings and graphite brushes to the outside of the machine. There, they can be short-circuited over rotor resistances. This can improve the behaviour of the asynchronous machine during start-up.

The most important advantage of the asynchronous machine is its simple and robust construction since a cage rotor has no sliprings, which could be mechanically strained.

The synchronous speed n_S of the rotating field in the stator can again be calculated from the mains frequency f_1 and the pole pair number p:

$$n_S = \frac{f_1}{p} \tag{5.88}$$

When the asynchronous machine is at a standstill, the rotating field of the stator passes the standing rotor. This induces a voltage in the conductors of the rotor windings. Due to this voltage, bar-type currents emerge in the bars of the winding in the rotor. The magnetic field of these currents creates a tangential force and the rotor starts to move. In motor mode, the rotor speed n is always lower than the synchronous speed n_S of the stator since a speed difference is needed to induce voltages in the rotor. The relatively small difference between rotor speed n and synchronous speed n_S is called *slip*:

$$s = \frac{n_S - n}{n_S} \tag{5.89}$$

If an asynchronous machine works as a generator, the rotor moves faster than the rotating field of the stator. The machine generates active power and feeds it via the stator to the mains. Table 5.7 summarizes different operating conditions of an asynchronous machine.

In contrast to the synchronous generator, the asynchronous generator always needs reactive currents as shown in the next section. An overexcited synchronous machine or power electronics can generate the necessary reactive

Table 5.7 *Speed and Slip at Different Operating Conditions for an Asynchronous Machine*

Operating condition	Speed	Slip
Standstill (short circuit)	$n = 0, n_S > 0$	$s = 1$
Motor mode	$0 < n < n_S$	$0 < s < 1$
Synchronous speed (no load)	$n = n_S$	$s = 0$
Generator mode	$n > n_S$	$s < 0$
Brake mode	$n < 0, n_S > 0$	$1 < s < \infty$

currents. If the asynchronous generator works in an island grid, a capacitor bank can generate reactive currents and provide the required *reactive power*.

Equivalent circuits and circle diagrams for the stator current

The electrical structure of an asynchronous machine is similar to a transformer that consists of two coupled windings. The windings of an ideal transformer are coupled free of leakages. A real transformer has leakage fields and ohmic losses that can be considered by additional resistances and reactances in the equivalent circuit (see Figure 5.27).

Hence, the equations for the voltages of the transformer become:

$$\underline{V}_1 = R_1 \cdot \underline{I}_1 + j X_{1\sigma} \cdot \underline{I}_1 + j X_{h1} \cdot \underline{I}_1 + j X_{12} \cdot \underline{I}_2 \tag{5.90}$$

$$\underline{V}_2 = R_2 \cdot \underline{I}_2 + j X_{2\sigma} \cdot \underline{I}_2 + j X_{h2} \cdot \underline{I}_2 + j X_{12} \cdot \underline{I}_1 \tag{5.91}$$

With the number of turns per phase w_1 and w_2 and $\underline{I}_2 = \frac{w_1}{w_2} \cdot \underline{I}'_2 = tr \cdot \underline{I}'_2$; $V_2 = V'_2 \, tr^{-1}$; $R_2 = R'_2 \cdot tr^{-2}$, $X_{2\sigma} = X'_{2\sigma} \cdot tr^{-2}$, $X_{h2} = X_h \cdot tr^{-2}$, $X_{12} = X_h \cdot tr^{-1}$, $X_{h1} = X_h$ as well as $X_{h1} = X_h$, the voltage equations related to the primary side become:

$$\underline{V}_1 = R_1 \cdot \underline{I}_1 + j X_{1\sigma} \cdot \underline{I}_1 + j X_h \cdot (\underline{I}_1 + \underline{I}_2') \tag{5.92}$$

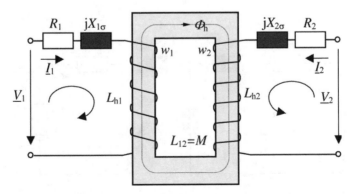

Figure 5.27 *Ideal Transformer with Resistances and Reactances*

$$\underline{V}'_2 = R'_2 \cdot \underline{I}'_2 + j X'_{2\sigma} \cdot \underline{I}'_2 + j X_h \cdot (\underline{I}_1 + \underline{I}'_2) \tag{5.93}$$

In contrast to the transformer, the rotor winding of an asynchronous machine, which corresponds to the secondary windings of the transformer, is short-circuited ($V'_2 = 0$). The rotor frequency f_2 of an asynchronous machine is different from the frequency f_1 of the stator. The slip s of the machine connects both frequencies:

$$\omega_2 = 2 \cdot \pi \cdot f_2 = s \cdot 2 \cdot \pi \cdot f_1 = s \cdot \omega_1 \tag{5.94}$$

The reactances of the rotor circuit become:

$$X = \omega_2 \cdot L = s \cdot \omega_1 \cdot L \tag{5.95}$$

If $X'_{2\sigma}$ and X_h in the voltage equations of the secondary site are replaced by $s \cdot X'_{2\sigma}$ and $s \cdot X_h$, respectively, and $V'_2 = 0$ is used, the equation for the rotor becomes:

$$0 = \frac{R'_2}{s} \cdot \underline{I}'_2 + j X'_{2\sigma} \cdot \underline{I}'_2 + j X_h \cdot (\underline{I}_1 + \underline{I}'_2) \tag{5.96}$$

The splitting

$$\frac{R'_2}{s} = R'_2 + R'_2 \cdot \frac{1-s}{s} \tag{5.97}$$

provides the *equivalent circuit* for one phase of the asynchronous machine (see Figure 5.28).

With $X'_2 = X'_{2\sigma} + X_h$ and the voltage equation of the rotor, the current becomes:

$$\underline{I}'_2 = -\frac{j X_h}{\dfrac{R'_2}{s} + j X'_2} \cdot \underline{I}_1 \tag{5.98}$$

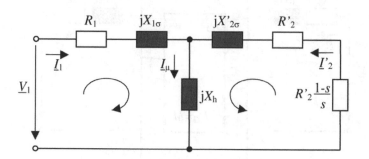

Figure 5.28 *Equivalent Circuit for One Phase of an Asynchronous Machine*

Substituting the current and $X_1 = X_{1\sigma} + X_h$ into the voltage equation of the stator and solving for I_1 provides:

$$\underline{I}_1 = \frac{\dfrac{R'_2}{s} + j X'_2}{(R_1 + j X_1) \cdot \left(\dfrac{R'_2}{s} + j X'_2\right) + X_h^2} \cdot \underline{V}_1 \tag{5.99}$$

This equation describes the *circle diagram* of the stator current. If the stator voltage $\underline{V}_1$ and the mains frequency f_1 as well as the resistances and reactances remain constant, the stator current depends only on the slip s.

Figure 5.29 shows the curve of the stator current $\underline{I}_1$ as a function of the slip. The stator voltage $\underline{V}_1$ is the reference value on the real axis. The current moves on a circle depending on the operating conditions of the machine. This circle is named after *Heyland and Ossanna*. The stator resistance R_1 is neglected for the Heyland circle. This circle clearly shows that the current at the stationary condition is much higher than that near zero-load operation.

With the major simplification $X_h \rightarrow \infty$, the current through X_h becomes zero. Figure 5.30 shows the simplified equivalent circuit with $R_1 \approx 0$ and $X_\sigma = X_{1\sigma} + X'_{2\sigma}$. This equivalent circuit is used for the derivation of the equation of Kloss in the section headed 'Speed–torque characteristics and typical generator data' on p223.

The simplified equivalent circuit provides a simplified equation of the current:

$$\underline{I}'_2 = -\underline{I}_1 = -\frac{\underline{V}_1}{\dfrac{R'_2}{s} + j X_\sigma} \tag{5.100}$$

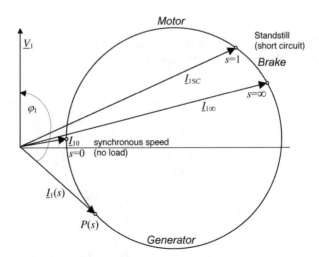

Figure 5.29 *Circle Diagram for the Estimation of the Stator Current According to Heyland and Ossanna*

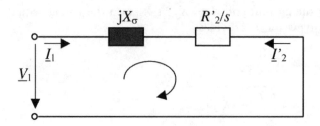

Figure 5.30 *Simplified One-phase Equivalent Circuit for an Asynchronous Machine*

Power balance

High generator efficiencies are very important for wind power plants because the majority of the mechanical power of the rotor blades connected to the generator should be converted to electrical power and fed into the grid. The ratio of the electrical power output P_1 to the mechanical power input P_2 defines the generator efficiency η. Asynchronous generators that are used in large wind generators can reach efficiencies over 95 per cent. Although the efficiencies are very good, the remaining losses are still very important. Losses in the range of 5 per cent at generator powers of more than 1000 MW produce large amounts of waste heat that must be removed. Figure 5.31 shows the power balance and the different losses of an asynchronous generator.

The available mechanical input power P_2 at the clutch is reduced by friction losses P_f due to mechanical and air friction. The rotor windings have ohmic losses, the so-called copper losses P_{Cu2}. The iron losses P_{Fe2} at the rotor can be neglected. The reduced air-gap power P_g is transferred from the rotor to the stator. It is given by:

$$P_g = P_2 - P_f - P_{Cu2} \tag{5.101}$$

The copper losses P_{Cu2} of the rotor winding occur at resistance R'_2 (see Figure 5.28). For three-phase windings these losses become:

$$P_{Cu2} = 3 \cdot I_2'^2 \cdot R'_2 \tag{5.102}$$

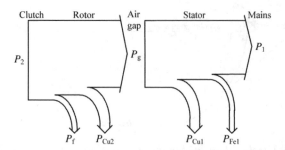

Figure 5.31 *Power Balance for an Asynchronous Generator*

The *air-gap power* P_g can be estimated by:

$$P_g = 3 \cdot I_2'^2 \cdot \frac{R_2'}{s} = \frac{P_2 + P_f}{1 - s} \tag{5.103}$$

Ohmic or copper losses in the stator

$$P_{Cu1} = 3 \cdot I_1^2 \cdot R_1 \tag{5.104}$$

and iron losses in the stator P_{Fe1} further reduce the air-gap power. The *output power*

$$P_1 = 3 \cdot V_1 \cdot I_1 \cdot \cos \varphi \tag{5.105}$$

is finally fed into the mains.

Speed–Torque Characteristics and Typical Generator Data

The connection between torque and speed for an asynchronous generator is also interesting because the rotor speed changes with the driving torque. The intrinsic torque of the machine is used for the following calculations. It is calculated from the air-gap power P_g and the synchronous speed n_S of the stator field as:

$$M_i = \frac{P_g}{2 \cdot \pi \cdot n_S} = \frac{3}{2 \cdot \pi \cdot n_S} \cdot \frac{R_2'}{s} \cdot I_2'^2 \tag{5.106}$$

The internal torque M_i is also given by the mechanical torque M (negative sign in generator mode) and the friction torque M_f (positive sign):

$$M_i = M + M_f = \frac{P_2}{2 \cdot \pi \cdot n} + \frac{P_f}{2 \cdot \pi \cdot n} \tag{5.107}$$

With the equation of the current for the simplified equivalent circuit:

$$I_2'^2 = \left| \underline{I}_2' \right|^2 = \frac{\left| \underline{V}_1 \right|^2}{\left| \dfrac{R_2'}{s} + j X_\sigma \right|^2} = \frac{V_1^2}{\dfrac{R_2'^2}{s^2} + X_\sigma^2}, \tag{5.108}$$

the internal torque becomes:

$$M_i = \frac{3}{2 \cdot \pi \cdot n_S} \cdot \frac{R_2'}{s} \cdot \frac{V_1^2}{\dfrac{R_2'^2}{s^2} + X_\sigma^2} = \frac{3 \cdot V_1^2}{4 \cdot \pi \cdot n_S \cdot X_\sigma} \cdot \frac{2}{\dfrac{R_2'}{s \cdot X_\sigma} + \dfrac{s \cdot X_\sigma}{R_2'}} \tag{5.109}$$

With $M_{iB} = \dfrac{3 \cdot V_1^2}{4 \cdot \pi \cdot n_S \cdot X_\sigma}$ \tag{5.110}

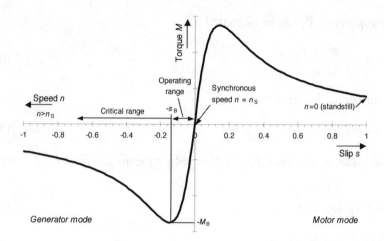

Figure 5.32 *Speed-torque Characteristics for an Asynchronous Machine*

and $\quad s_B = \dfrac{R'_2}{X_\sigma},$ (5.111)

the so-called *Kloss formula* can be obtained:

$$M_i = M_{iB} \frac{2}{\dfrac{s}{s_B} + \dfrac{s_B}{s}}$$ (5.112)

If the slip s is equal to the value s_B, the second term becomes unity. The internal torque M_i reaches a maximum at this point. This maximum value is also called the *breakdown torque* M_{iB}. The slip at this torque is called the *breakdown slip* s_B.

Figure 5.32 shows the speed-torque characteristics for an asynchronous machine. When starting-up, an asynchronous generator operates in motor mode starting at the slip $s = 1$. After passing the synchronous operating point of $s = 0$, the generator reaches the normal operating range. If the load at the generator becomes higher than the breakdown torque M_B, the speed increases rapidly. Then, the generator can overspeed and destroy the system mechanically. The normal operating range of an asynchronous generator is between 0 and $-s_{B'}$. The speed is a little higher than the synchronous speed n_S.

Table 5.8 shows data for a real asynchronous generator used in a wind generator. The rated slip (that is, the relative speed difference between rotor and stator) is $-0.05 = -5$ per cent. The normal speed range is between 1515 min^{-1} and 1650 min^{-1}, which can be obtained from the pole pair number p = 2 and the frequency of the mains f_1 by:

$$n = (1-s) \cdot \frac{f_1}{p} \cdot \tfrac{60s}{min}$$ (5.113)

The power factor $\cos\varphi$ is 0.88, far from the ideal value $\cos\varphi = 1$. During part-load operation, $\cos\varphi$ is even worse and reaches values around 0.5 at 25 per cent part load. This causes a high amount of reactive power that must be reduced by compensation circuits. A capacitor bank for instance can make this compensation and achieve $\cos\varphi$ factors of about 0.99.

ELECTRICAL SYSTEM CONCEPTS

Asynchronous generator with direct mains coupling

A very simple wind generator concept is the so-called *Danish concept*. It is mainly used for small- and medium-sized wind generators that have been developed in Denmark. These systems connect a stall-regulated asynchronous generator directly to the mains (see Figure 5.33). A gearbox adjusts the speed of the rotor blades to the generator speed. This system concept appeals with its simplicity. The asynchronous generator need not be synchronized with the grid as with a synchronous generator. It reaches its operating speed without further control. However, very large generators can cause high starting currents when they are connected to the mains. Controlled-torque starting circuits can be used to limit these currents.

The stall regulation of the rotor limits the power at high wind speeds. The generator can cushion rapid fluctuations in the wind speed while it changes its speed through the slip s. Asynchronous wind generators allow changes in the speed of the order of about 10 per cent. However, losses get higher and efficiencies worse if the slip rises. Therefore, modern concepts use asynchronous machines with *variable slip*. These generators have no cage rotor with short-circuited ends of the windings, but controllable resistances R_R in the rotor circuit instead. The rotor windings are either connected via slip rings to controllable resistances outside the machine or to controllable resistances rotating with the rotor. Figure 5.34 shows the change of the speed-torque characteristics for asynchronous machines if resistances are connected to the rotor circuit. The breakdown torque moves to higher slip values for higher rotor resistances. Since the power is proportional to the torque, the speed increases at higher powers and cushions power fluctuations.

A speed-power diagram shows how well suited an asynchronous generator is for connection to the mains.

Table 5.8 *Technical Data for a 600-kW Asynchronous Wind Generator*

Nominal power P_N	600 kW	Reactive power Q at full load	324 kvar
Rated voltage V_N	690 V	$\cos\varphi$ at full load	0.88
Mains frequency f_1	50 Hz	Speed range n	1515–1650 min⁻¹
Rated current I_N	571 A	Nominal speed n_N	1575 min⁻¹
Pole number $2\cdot p$	4	Slip range s	−0.1–0
Winding connection	Star connection	Nominal efficiency η_N	95.2%

Source: Vestas, 1997

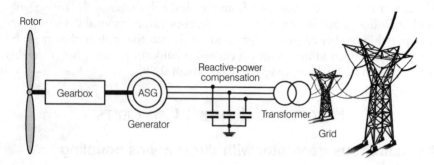

Figure 5.33 *Asynchronous Generator with Direct Mains Coupling*

With $\lambda = \dfrac{2 \cdot \pi \cdot r}{v} \cdot n$, the approximation for c_p of a certain wind generator is:

$$c_p = 0.00068 \cdot \lambda^3 - 0.0297 \cdot \lambda^2 + 0.3531 \cdot \lambda - 0.7905 \quad \text{(see also Equation 5.45)}$$

and with $P = c_p \cdot P_0 = c_p \cdot \frac{1}{2} \cdot \rho \cdot A \cdot v^3$,

the dependence of rotor power P on the speed n for a constant wind speed v can be calculated (see Figure 5.35). However, the asynchronous generator allows only relatively small fluctuations in speed. The stator frequency and the gearbox define the rotor speed, and this can only be varied by the slip that increases with the power. Asynchronous generators with variable slip can increase the slip at higher powers. This can cushion high power fluctuations and high strains to the mains.

Figure 5.35 shows also that the rotor speed has an important influence on the usable wind energy. If the rotor speed is nearly constant, the wind generator

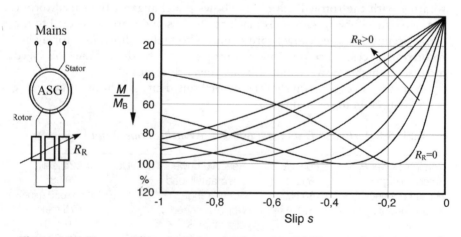

Figure 5.34 *Torque Characteristics as a Function of Slip* s *with Variation of the Rotor Resistance* R_R

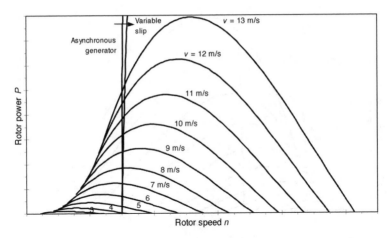

Figure 5.35 *Operating Points for a Wind Turbine with Asynchronous Generator that is Directly Coupled to the Mains*

does not use the optimal power at all wind speeds. For this example, no power can be taken from the wind at wind speeds below 4 m/s because the rotor speed is too high. Maximum power usage is achieved at wind speeds of about 8 m/s. The relative share of the power that can be taken from the wind decreases with rising wind speeds.

System concepts with two different rotational speeds achieve higher energy gains. One concept uses two asynchronous generators that can be coupled to the rotor one after the other. Another concept applies one asynchronous *change-pole generator*. The stator of this generator has two separate windings with different numbers of poles. When switching from one winding to the other the stator speed also changes because the rotor speed directly depends on the pole pair number. Hence, systems that work over two speed ranges can be designed. Figure 5.36 shows that the ranges for taking the optimum power from the wind can be extended by two different generator speeds. The first asynchronous generator of this example operates at wind speeds between 3 m/s and 7 m/s, the second at wind speeds above 7 m/s.

The main disadvantage of the asynchronous machine is the demand for reactive current. If the asynchronous machine is connected to the mains, the mains can provide this reactive current. However, electric utilities usually want high compensatory payments so that it is more economical to install *reactive power compensators*. This can be, for instance, a capacitor bank as shown in Figure 5.33. This capacitor bank can usually be designed for some operating points only, so that the mains must still provide the remaining reactive power. Modern power electronics can better compensate the reactive power.

The conditions for island systems are totally different. Since the reactive power demand rises with increasing power, it must be controlled within the island system. Power electronic units or a synchronous machine, which operates as a phase modifier, can provide the respective reactive power demand.

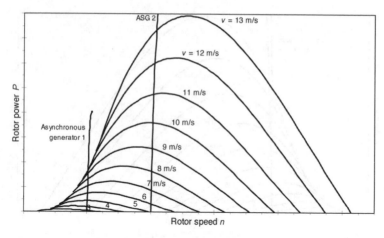

Figure 5.36 *Operating Points for a Wind Turbine with Two Asynchronous Generators with Different Speeds*

Synchronous generator with direct mains coupling

The synchronous generator does not have the disadvantage of requiring a certain reactive power demand since this can be controlled by the excitation current. Therefore, a synchronous generator can both generate and consume reactive power. Permanent magnets can also be used to excite the synchronous generator. In this case, reactive power cannot be controlled. Furthermore, permanent magnets are rather expensive. Hence, power electronic components such as thyristors usually convert the three-phase current of the mains to the desired DC that determines the excitation (see Figure 5.37).

In contrast to the asynchronous generator, the synchronous generator runs with a constant speed. The operation characteristics in the speed-power diagram are indicated by an absolutely vertical line that has no slight bend at higher powers as with the asynchronous generator. The slip cannot cushion power jumps so that they pass almost entirely to the mains. Besides changes in load at the mains, this can cause high mechanical strains on the wind generator itself. A slip clutch can cushion gusts but also exhibits high wear.

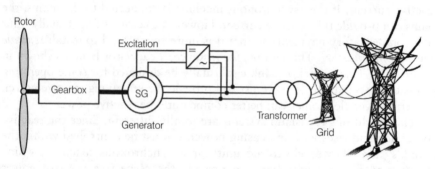

Figure 5.37 *Synchronous Generator with Direct Mains Coupling*

Therefore, synchronous generators are usually not coupled directly to the mains as shown in Figure 5.37. The concept of direct coupling is mainly used for stand-alone systems in island grids. In such grids, usage may consist of pumps driven by three-phase motors or DC loads with rectifiers and batteries. This allows the synchronous generator to work at different operating points with different speeds, which reduces the loads significantly.

Synchronous generator with DC-link converter

Modern power electronics can avoid the disadvantages of a synchronous generator that is directly coupled to the mains. The synchronous generator is connected via a DC link or frequency converter to the mains (see Figure 5.38). In this configuration, the system can operate the generator at a frequency independent of the mains frequency. Changing the generator frequency varies the generator speed. Hence it is possible to vary the speed over a wide range and to run at the optimal speed to obtain the maximum power use depending on the wind speed.

Figure 5.39 shows clearly that it is possible to take the maximum power from the wind at low and medium wind speeds by changing the rotor speed. At high wind speeds it is necessary to limit the power. Two possibilities exist. In the first solution, the power electronics keeps the frequency and therefore the speed constant. Then a stall control system can limit the power (see also Figure 5.14). In the second possibility, an inverter limits the power at high speeds.

However, neither solution can completely avoid the risk of rotor overspeeding at very high wind speeds; therefore, an additional power limiting mechanism is used, for instance by adjusting the pitch of the rotor blades. This shifts the rotor curves to smaller speeds than shown in Figure 5.39.

Since the frequency converter can achieve other rotor frequencies than that given by the mains, a gearbox for adjusting the rotor speed to the generator speed is no longer needed. Gearless wind turbines are produced in high numbers even in the megawatt class today. Generators for these turbines are high-pole synchronous generators with 80 or more poles. Besides the reduced material requirements and resulting cost reductions, *gearless wind power*

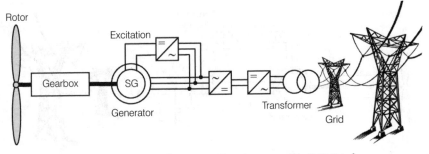

Figure 5.38 *Synchronous Generator with DC Link*

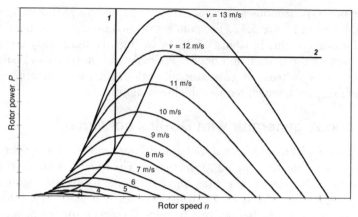

Figure 5.39 *Operating Points for a Variable-Speed Wind Generator with Power Limited by Constant Speed (1) or by a Converter (2)*

plants have other advantages such as lower noise levels. If pulse width modulated inverters are used, the power electronics can also control the reactive power demand.

Speed-controlled asynchronous generators

For directly connected asynchronous generators it has already been shown that a variable slip could change the speed of the generator. Since high slips cause high losses in the rotor circuit, the slip was limited to 10 per cent in the examples above. However, it is also possible to use the rotor power and feed it into the mains (see Figure 5.40). If there is only the possibility to vary the generator speed above the mains frequency, the circuit is called an *oversynchronous converter cascade*. A disadvantage of this circuit is the high reactive power demand.

Whereas an oversynchronous converter cascade can only transfer power from the rotor to the mains, a *double-fed asynchronous generator* can also transfer power from the mains to the rotor. Therefore, a DC link converter or a direct converter can be used (see Figure 5.41).

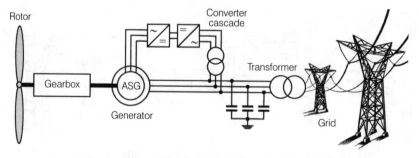

Figure 5.40 *Variable Speed Asynchronous Generator with Converter Cascade*

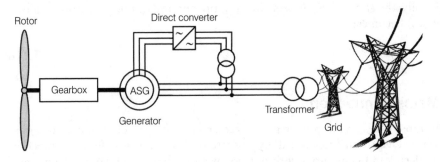

Figure 5.41 *Double-fed Asynchronous Generator with Direct Converter*

The double-fed asynchronous generator can be operated both oversynchronously and subsynchronously. In other words, rotor speeds below the synchronous mains frequency are possible. Hence, the reactive power demand of the generator can be controlled. Disadvantages of this system are mains pollution and higher costs. Because of the possibility of the change in speed and the controllable reactive power demand, this system has nearly the same advantages as a gearless synchronous generator with converter, so that it is increasingly being used.

MAINS OPERATION

Annual energy gain

Before installing a wind generator, an annual performance calculation must be performed. Therefore, the annual system output must be calculated. On-site measurements of the wind speed usually provide the annual wind speed distribution. For large wind generators or for wind farms, several wind surveys are often made. With the manufacturer's figures for the dependence of the power $P(v)$ on the wind speed v (see also Figure 5.13) and the wind speed distribution $f(v)$, the average annual power $\overline{P}$ becomes:

$$\overline{P} = \int_0^\infty f(v) \cdot P(v) \cdot \mathrm{d}v \qquad (5.114)$$

The Weibull or Rayleigh distributions, covered in the previous section on wind speed distributions, are often used to describe the wind speed distribution. If on-site measurements are available, relative frequency distributions $h(v)$ with different wind speed intervals i as shown in Figure 5.1 are given. Hence, using the average wind generator power $P(v_i)$ of the respective interval i, the annual average power becomes:

$$\overline{P} = \sum h(v_i) \cdot P(v_i) \qquad (5.115)$$

Finally, the *energy gain E* over time *t* (a time period of one year is used in most cases) becomes:

$$E = \overline{P} \cdot t \tag{5.116}$$

Mains connection

In most cases, wind power plants are erected in sparsely populated regions where electric utilities normally operate only weak grids. Before connecting modern wind generators with powers of more than 1000 kW to the mains, it must be checked whether the mains has the capacity to take the generated currents. To keep the transmission losses low, high voltages are chosen for the transmission of high powers. The voltage ranges are subdivided into:

*   low voltage   0.1–1 kV
*   medium voltage   1–35 kV
*   high voltage   35–230 kV.

Usually wind power stations are connected via a transformer to the medium- or high-voltage system.

Since both the mains and the wind power station can cause technical problems, protection equipment must disconnect the wind generator from the mains in case of error detection, thereby protecting the mains voltage and frequency. The protection equipment must switch before there is a voltage rise in one phase of 15 per cent, a voltage drop of 30 per cent or a deviation of the frequency from the nominal frequency of 2 Hz.

Connecting generators to the mains also produces voltage fluctuations that should not be greater than 2 per cent. The *maximum rated power* S_{rG} of the wind generator is obtained from the short-circuit power S_{SC} at the point of common coupling and the factor *k* as given in Table 5.9 (VDEW, 1994):

$$S_{rG} \leq 0.02 \cdot \frac{S_{SC}}{k} \tag{5.117}$$

Hence, it is necessary to know the short-circuit power of the mains to calculate the permissible generator power.

Another aspect that must be considered when connecting a wind generator to the mains is so-called *flicker*. A series of voltage fluctuations can cause changes in the luminance of light bulbs. If the generator power is lower than a thousandth of the short-circuit power of the mains, no further investigations are necessary. At higher powers, disturbing luminance fluctuations are possible. To avoid these, a so-called long-term disturbing flicker factor A_{lt} is estimated. With the relative voltage change *d* (for instance $d = S_{rG}/S_{SC}$) and the form factor *F* (*F* = 1 for one voltage jump) the flicker after-effect time of one voltage change becomes:

Table 5.9 *Values of the k Factor for the Calculation of the Rates of Generator Power*

Synchronous generators, inverter	$k = 1$
Asynchronous generators, powered up with 95–105% of their synchronous speed	$k = 2$
Asynchronous generators, run up by the mains	$k = I_l/I_r$
If I_l is unknown	$k = 8$

Note: I_l locked-rotor current, I_r rated current

$$t_f = 2{,}3\,\text{s} \cdot (100 \cdot d \cdot F)^3 \tag{5.118}$$

The sum of the flicker after-effect times over an investigation interval of 120 min provides the long-term disturbing flicker factor:

$$A_{lt} = \frac{\sum t_f}{120 \cdot 60\,\text{s}} \tag{5.119}$$

For wind generators, the system flicker coefficient c can be estimated from measurements (VDEW, 1994). This provides the criterion for the permissible long-term disturbing flicker coefficient:

$$A_{lt} \leq \left(c \cdot \frac{S_{rG}}{S_{SC}} \right)^3 \tag{5.120}$$

A boundary criterion for the long-term disturbing flicker factor A_{lt} of a single wind generator is:

$$A_{lt} \leq 0.1 \tag{5.121}$$

The higher the number of rotor blades the lower is the system flicker coefficient c. If the generator is connected via an inverter to the mains, the flicker coefficient is usually smaller than that of systems connected directly to the mains. If multiple wind generators of a wind park are connected to the same knot of the medium- or high-voltage system, the resulting system flicker coefficient c_{res} of n similar generators becomes:

$$c_{res} = \frac{c}{\sqrt{n}} \tag{5.122}$$

The problem of harmonics have already been discussed in detail in the chapter on photovoltaics (Chapter 4). Since the current of wind generators is usually fed into medium- or high-voltage systems, the limiting values are different from those of the low-voltage systems to which photovoltaic systems are usually connected.

If one of the guide values is exceeded, technical measures must be put in place to mitigate them. If the short-circuit power of the mains is too low, another mains connecting point with lower impedance and higher short-circuit power must be found. This also increases the permissible harmonic currents. However, in most cases this solution needs new, expensive transmission lines. Harmonics can be also eliminated with harmonic suppressors or highly pulsed pulse width modulated inverters. Voltage fluctuations and flickers can be reduced by a regulation of the power flow. Therefore, wind generators with variable-speed or asynchronous generators with variable slip are advantageous, as described above. Short-term storage systems, which can buffer high power fluctuations, can provide further improvements (Carstens and Stiebler, 1997). Battery storage systems, flywheel storage systems, high-power capacitors or superconducting coils can therefore be used.

Chapter 6

Economics

INTRODUCTION

The question of economic efficiency is a major concern when planning renewable power projects. Alleged poor economic efficiency is one of the main arguments against renewable energy sources. The solution with the best economic benefits is usually the one realized, whereas technical or ecological aspects are of secondary importance. From the national economy's point of view, accommodating these latter considerations often causes negative consequences. Finally, effects on the environment are not considered sufficiently in current practice. Therefore, the second part of this chapter gives a critical analysis of common methods of valuation.

All the systems described in this book are technical systems for energy conversion. The aim of economic calculations is to find the one system out of the various possible solutions that provides the desired type of energy at the lowest cost. Therefore, different types or variants of renewable energy systems are compared one against the other. Furthermore, renewable energy systems are usually compared with conventional systems, although many such comparisons do not consider the full benefits and costs for the national or world economy.

The result of economic calculations is the *cost for one unit of energy*. For heat-providing systems the costs are related to a kilowatt-hour of heat ($€/kWh_{therm}$). For electricity-generating systems the costs are related to a kilowatt-hour of electricity ($€/kWh_{el}$). For estimating specific final cost, all the costs such as installation of the power plant, operating and maintenance costs as well as disposal costs are divided by the total number of kilowatt-hours generated during the plant's lifetime. For the estimation of costs, many assumptions about future developments have to be made. In many cases, reality is different from the predictions and this fact may change real costs significantly. Often it is the taxpayer, i.e. the general public, who must pay for these miscalculations. In these cases, certain costs are sometimes excluded from the economic calculations so that the power plant appears profitable again. Examples include the underestimation of costs for the shutdown and disposal of nuclear power stations, ongoing inspection of final disposal sites or restoration of exploited opencast coal mining sites.

When comparing costs, prices from different dates are often given (e.g. over the lifetime of a plant). In this case *inflation* must be considered. Therefore, it is necessary to always indicate the reference year for the cost and

Table 6.1 *Consumer Price Index (CPI) for the US, Reference Year 1967 (1967 = 100)*

Year	1800	1825	1850	1875	1900	1910	1920	1930	1940	1945	1950	1955
CPI	51	34	25	33	25	28	60	50	42	53.9	72.1	80.2

Year	1960	1965	**1967**	1970	1975	1980	1982	1984	1986	1988	1990	1991
CPI	88.7	94.5	**100.0**	116.3	161.2	246.8	289.1	311.1	328.4	354.3	391.4	408.0

Year	1992	1993	1994	1995	1996	1997	1998	1999	2000	2001	2002	2003
CPI	420.3	432.7	440.0	456.5	469.9	480.8	488.3	499.0	516.0	530.4	538.8	550.3

Source: Federal Reserve Bank of Minneapolis, 2003

price estimation (e.g. $€_{1995}$/kWh). Table 6.1 shows the consumer price index for the US over the past two centuries. However, price increase may be different for other countries. Besides general price increases, prices of conventional energy resources can vary significantly, as the oil crises of the 1970s have shown beyond any possible doubt. Therefore, cost calculations for conventional energy systems have relatively high uncertainty and high economic risks, which are often ignored. Since renewable energy sources such as solar and wind energy are free, no change in fuel prices can affect the costs of renewable energy systems.

CLASSICAL ECONOMIC CALCULATIONS

Calculations without return on capital

In classical economic calculations the investor expects a payment of interest for his invested capital. The interest rate depends mainly on the risk, followed by other factors. However, in this section, economic calculation reference values are estimated without return on capital (see also the section headed 'Critical View of Economic Calculations', p254).

The estimation of the cost for one unit of energy without considering return on capital is relatively easy. All costs over the whole lifetime of the energy system are added and then divided by the operating period. The result is the total annual cost. Dividing this cost by the annual generated units of energy of the system provides the cost per unit of energy.

The *total cost* C_{tot} consists of the *investment cost*, i.e. payment A_0 for year zero for the system installation, and payments A_i for every operational year i, respectively, for n years. *Operating cost* could be the cost of maintenance, repair, insurance and for some systems, fuel costs. These costs may vary from year to year. For an operating period of n years, the total cost becomes:

$$C_{tot} = A_0 + \sum_{i=1}^{n} A_i \qquad (6.1)$$

Table 6.2 *Breakdown of the Costs of Grid-connected Photovoltaic Systems*

System size	1 kWp (%)	5 kWp (%)
Photovoltaic module	48	55
Inverter	13	13
Installation material	17	16
Installation (labour costs)	15	11
Planning / documentation	7	5

Source: data from Becker and Kiefer, 2000

Hence, the average total annual cost C_a becomes:

$$C_a = \frac{C_{tot}}{n} \tag{6.2}$$

Assuming that the system produces an average annual amount of energy E_a the *specific energy cost* C_E is calculated by:

$$C_E = \frac{C_a}{E_a} \tag{6.3}$$

This cost is also called the levelled electricity cost (LEC) or levelled heat cost (LHC).

Costs for a photovoltaic system

Whereas average investment costs of *grid-connected photovoltaic systems* exceeded €10,000/kWh at the beginning of the 1990s, these costs have been falling significantly. The cost reduction for photovoltaic systems has been nearly 50 per cent per decade over the past two decades. End user prices for single photovoltaic modules are now below €5/W_p. Inverters, support structures, cables and installation result in further costs. The specific investment costs decrease with the system size. The end user prices for grid-connected photovoltaic systems in the range of 5 kW_p were about €7500/kW_p in 2001, whereas prices close to €5000/kW_p can be achieved for systems in the megawatt size range. Table 6.2 shows that the photovoltaic modules make up only about half of the total system cost.

The cost of one kilowatt hour of electricity generated by a photovoltaic system is calculated here as an example. The underlying numbers have been also used in the examples later in the chapter. The investment costs for a system with a rated power of 1 kW_p are assumed to be A_0 = €6500. The operating period of the system is 25 years (n = 25) and the system generates E_a = 800 kWh of electricity per year. This generation level can be expected in regions with an annual solar irradiation of about 1000 kWh/m^2, as found in central Europe. In the tenth year of operation, one system component, for instance the inverter, is likely to need replacing. This cost therefore is A_{10} = €1500.

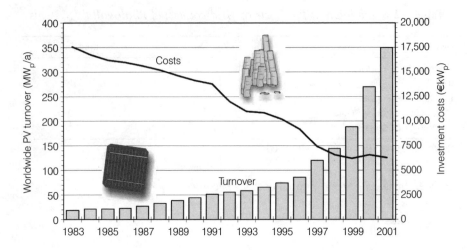

Figure 6.1 *Global Photovoltaic Module Production and End User Prices for Small Grid-connected Photovoltaic Systems in Germany*

Further costs should not accrue in this example. Thus, the levelled electricity cost is calculated as follows:

$$C_{tot} = A_0 + A_{10} = €8000$$

$$C_a = C_{tot}/n = €320 \text{ per year}$$

$$C_E = C_a/E_a = €0.40/\text{kWh}_{el}.$$

In regions with very high annual irradiance, for example, near the Sahara desert, the same system can generate up to 2000 kWh/year of electricity. This reduces the levelled electricity cost without return on capital from €0.40/kWh$_{el}$ to €0.16/kWh$_{el}$. The cost for very large systems is even lower. Today, these costs are usually higher than the conventional electricity price. However, they will remain relatively constant throughout the operating period since no high uncertainty about fuel prices exist.

Future photovoltaic systems will achieve further significant cost reduction as a result of the fast increasing production volume. In 2002, the amount of photovoltaic modules produced annually worldwide was about 560 MW, which is similar to the amount produced by a very small conventional power plant, but the annual increase in photovoltaic module production is consistently more than 20 per cent.

Figure 6.1 shows that costs decrease significantly with rising production volume. Since the mid-1980s, costs have decreased by more than 60 per cent.

Costs for a wind power plant

The investment costs of wind power plants have also decreased during the past

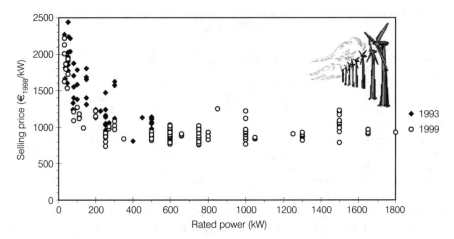

Source: data from Wind-Kraft Journal, 1993, 1999

Figure 6.2 *Specific Sale Prices for Wind Turbines in 1993 and 1999*

few years. On one hand, costs for wind power plants with the same capacity have decreased. On the other hand, the rated power of wind turbines has grown rapidly during the same time. This reduces the incidental expenses per kilowatt. Figure 6.2 shows the sale prices and rated powers of wind turbines in two different years. However, the price conditions vary between different turbine designs.

Besides the pure cost of the wind turbine, investment costs include costs for planning, installation, foundations, mains connection and transport. These costs are only partly included in the values of Figure 6.2. These incidental expenses are on average 34.5 per cent of the wind turbine sale price, but can be much higher for smaller systems. Operating costs consists of land rent, maintenance, repair and insurance costs.

The total investment cost for a wind turbine in the 1.5 MW range is currently between €1000/kW and €1600/kW depending on the site and the required infrastructure expenses. Using €1200/kW for a 1.5-MW turbine, the investment cost A_0 is €1,800,000. The annual operating costs are usually between 2 and 3 per cent of the investment costs, so annual operating costs A_i of about €50,000 must be considered.

A 1.5 MW turbine at a site with a wind speed of just under 7 m/s at hub height, corresponding to 4.5 m/s at 10 m, can generate about E_a = 3.5 million kWh per year. Assuming an operating period of 20 years for the wind turbine, the calculations become:

$$C_{tot} = A_0 + 20 \cdot A_i = €1,800,000 + 20 \cdot €50,000 = €2,800,000$$

$$C_a = C_{tot}/n = €140,000$$

$$C_E = C_a/E_a = €0.04/kWh_{el}$$

Table 6.3 *Annual Energy Gain for Wind Power Plants of Different Sizes and Different Wind Speeds* v_{hub}

System size		Annual energy gain in MWh$_e$/a				
Rotor	Power (kW)	v_{hub} = 5 m/s	6 m/s	7 m/s	8 m/s	9 m/s
30 m	200	320	500	670	820	950
40 m	500	610	970	1360	1730	2050
55 m	1000	1150	1840	2570	3280	3920
65 m	1500	1520	2600	3750	4860	5860
80 m	2500	2380	4030	5830	7600	9220
120 m	5000	5300	9000	13,000	17,000	20,000

These costs, excluding return on capital, are close to the costs of conventional power plants. Renewable energy sources legislative acts in Germany and Spain guarantee a payment for wind-generated electricity of above €0.06/kWh. This has made wind power competitive at many sites without any further subsidies and initiated the present wind power boom.

However, at average sites with a wind speed below 5 m/s at 10 m height, wind turbines can normally be operated economically only with grants since the annual energy gain decreases significantly as shown in Table 6.3. The table also shows that the energy gain relative to the power rating also increases with the turbine size. The higher hub height of large systems is the main reason behind this. Theoretically, small turbines can also be installed with higher hub heights but this usually makes the economics worse. Larger system sizes also offer the potential for future cost reduction.

Costs for a solar thermal system for domestic water heating

The costs for solar domestic water heating systems break down as just under a third for the solar collector, just over a third for the hot water storage tank and accessories and a third for the installation. In industrialized countries the prices of flat-plate collectors are between €200/m² and €350/m² and prices of evacuated tube collectors are in the range of €700/m². A 300 litre stainless steel water tank costs between €700 and €1100. Hence, the price of a reasonable system with 4 m² of flat-plate collector and a 300 litre hot water tank without installation is about €2700. The average cost for a collector system with forced circulation used for a single household is between €3000 and €3500 without installation and about €5000 including installation. If the solar energy system is installed in a new building or if the conventional heating system must be renewed anyway, the cost for the system can be lower. Public subsidies can also decrease the costs. However, due to much lower labour costs and simpler materials used for the water tank and pipes, solar collector systems are considerably cheaper in countries such as China, Greece and Turkey. In China, a standard collector system with a 2 m² vacuum tube collector and a 180 litre water tank is sold for about €430 (ESTIF, 2003).

A common domestic solar water heating system is the basis for the following calculations. The annual heat demand in this example is 4000

Table 6.4 *Levelled Heat Costs in €/kWh$_{therm}$ for Solar Thermal Systems for Domestic Water Heating without Return on Capital and with Annual Operating Costs of €40*

Substituted amount of energy Lifetime	Annual 2300 kWh$_{therm}$		Annual 1150 kWh$_{therm}$	
	20 years	15 years	20 years	15 years
Investment costs				
€5000	0.13	0.16	0.25	0.32
€3500	0.09	0.12	0.19	0.24
€2500	0.07	0.09	0.14	0.18
€1800	0.06	0.07	0.11	0.14
€1000	0.04	0.05	0.08	0.09

kWh$_{therm}$ and the solar fraction 50 per cent. Hence 2000 kWh$_{therm}$ of useful heat can be substituted by the solar energy system each year. Assuming an average efficiency for the conventional reference system of 85 per cent, the annual substituted amount of conventional energy is 2353 kWh$_{therm}$. This amount of energy is used in the following calculations as the annually produced energy. However, if the solar energy system is installed in the Earth's sunbelt region, the solar fraction can be up to 100 per cent and the substituted energy much higher. In contrast, if the heat demand is lower, the substituted energy decreases as well. Relatively low annual maintenance costs of €31 are assumed and the pump of the solar cycle of this example requires 60 kWh$_{el}$ of electricity at a price of €0.15/kWh$_{el}$ (€9 per year). For an investment cost A_0 of €3300, annual operating costs A_i of €40 and an operating period of $n = 20$ years, the calculations become:

$$C_{tot} = A_0 + 20 \cdot A_i = €3300 + 20 \cdot €40 = €4100$$

$$C_a = C_{tot}/n = €205$$

$$C_E = C_a/E_a = €205/2353 \text{ kWh}_{therm} = €0.087/\text{kWh}_{therm}$$

Table 6.4 shows the calculations for different investment costs, operating periods and substituted amount of energy. In moderate climates such as those of central Europe, the solar heat generation costs are often above the costs of conventional heating systems. Solar pool heating systems are usually competitive due to much lower investment costs. The relative investment costs of large solar thermal systems and solar district heating systems are also lower so that they can operate at lower cost.

Calculations with return on capital

In addition to the calculations described above without return on capital, investors usually expect payments of interest for their investment of the order of usual paid rates for comparable investments. Starting with the payment A_0

in the initial year, the capital c_n after a time of n years, with the interest rate *ir*, can be estimated using compound interest formulae. With

$$q = 1 + ir,$$ (6.4)

the expression becomes:

$$c_n = A_0 \cdot (1 + ir)^n = A_0 \cdot q^n$$ (6.5)

For an initial capital of A_0 = €6500 (following the example of the photovoltaic system in the section headed 'Costs for a photovoltaic system', p237) invested with an interest rate of *ir* = 6 per cent = 0.06 over a time of n = 25 years, the capital after 25 years including interest becomes

$$c_{25} = €6500 \cdot (1 + 0.06)^{25} = €27,897$$

Later payments yield interest as well, though over a lower term. If the investor pays another €1500 after 10 years (for a replacement inverter), the term reduces to 15 years. The capital after 25 years would then be:

$$c_{25} = A_0 \cdot q^{25} + A_{10} \cdot q^{15} = €6500 \cdot 1.06^{25} + €1500 \cdot 1.06^{15} = €31,492$$

The following formula describes this in general:

$$c_n = A_0 \cdot q^n + \sum_{i=1}^{n} A_i \cdot q^{n-i}$$ (6.6)

If all payments A_i in the different years *i* are the same, the geometric progression provides the following formula:

$$c_n = A_0 \cdot q^n + \sum_{i=1}^{n} A \cdot q^{n-i} = A_0 \cdot q^n + A \cdot \frac{q^n - 1}{q - 1}$$ (6.7)

For later payments it is necessary to calculate the additional capital that would have to be invested at the beginning with the interest rate *ir* so that the capital at the end of the investment period is the same.

The payment A_i in the year *i* can be discounted to the initial year:

$$A_{i/0} = \frac{A_i}{q^i} = A_i \cdot q^{-i}$$ (6.8)

Thus, the respective initial capital for a payment of €1500 after 10 years for the example above becomes:

$$A_{10/0} = €1500 \cdot 1.06^{-10} = €838$$

In other words, if 6 per cent interest is paid on €838 over 25 years, the final capital is the same as if €1500 was invested after 10 years, bearing 6 per cent interest over 15 years.

The following equation provides the *discounting* of several such payments at different times:

$$c_0 = A_0 + \sum_{i=1}^{n} \frac{A_i}{q^i} \tag{6.9}$$

If the payments A_i in the different years i are all the same, it becomes:

$$c_0 = A_0 + A \cdot \frac{q^n - 1}{(q-1) \cdot q^n} \tag{6.10}$$

Then, the capital after n years can be calculated again with the equation for compound interest:

$$c_n = c_0 \cdot q^n \tag{6.11}$$

When investing in technical systems for energy conversion, there is likely to be little capital left that can be repaid after the end of its operating life in n years. On the contrary, most end-of-life systems are in a poor state of repair and therefore worthless. Selling the converted energy of the system yields income for the repayments of the invested capital during the operating time. Hence, it is now possible to calculate the price at which a unit of energy must be sold so that the investor gets the required rate of interest.

The sales return also yields interest. If the investor gets a repayment as early as the beginning of the first operating year, he can reinvest this amount with payments of interest over the whole operating time. The interest period decreases for later repayments. The initial capital c_0 is calculated as described above with the initial payment A_0 and the payments A_i in the following years discounted to the initial year. These payments are compared with the income B_i. They must also be discounted to the initial year. For simplification, payments and income within a year are treated as if they were made at the end of the year. For an operating period of n years the calculations become:

$$NPV = -A_0 + \sum_{i=1}^{n} \frac{B_i - A_i}{q^i} = -A_0 - \sum_{i=1}^{n} \frac{A_i}{q^i} + \sum_{i=1}^{n} \frac{B_i}{q^i} = -c_0 + \sum_{i=1}^{n} \frac{B_i}{q^i} . \tag{6.12}$$

NPV is called the *net present value*; it must be greater than or equal to zero if the investment is not to yield losses.

In the following, it is assumed that the income B at the end of each year is the same. The size of the required annual income can be estimated if the equation of the net present value is solved for B and the NPV is set to zero. With the annuity factor, a, and:

Table 6.5 *Annuity Factors a for Various Interest Rates ir and Interest Periods n*

n	Interest rate (discount rate) ir = q – 1									
	1%	2%	3%	4%	5%	6%	7%	8%	9%	10%
10	0.1056	0.1113	0.1172	0.1233	0.1295	0.1359	0.1424	0.1490	0.1558	0.1627
15	0.0721	0.0778	0.0838	0.0899	0.0963	0.1030	0.1098	0.1168	0.1241	0.1315
20	0.0554	0.0612	0.0672	0.0736	0.0802	0.0872	0.0944	0.1019	0.1095	0.1175
25	0.0454	0.0512	0.0574	0.0640	0.0710	0.0782	0.0858	0.0937	0.1018	0.1102
30	0.0387	0.0446	0.0510	0.0578	0.0651	0.0726	0.0806	0.0888	0.0973	0.1061

$$0 = -c_0 + B \cdot \sum_{i=1}^{n} \frac{1}{q^i} = -c_0 + B \cdot \frac{q^n - 1}{(q-1) \cdot q^n} = -c_0 + B \cdot \frac{1}{a} \quad \text{the result is}$$

$$B = c_0 \cdot a \tag{6.13}$$

The annuity factor a is:

$$a = \frac{q^n \cdot (q-1)}{q^n - 1} = \frac{q-1}{1 - q^{-n}} \tag{6.14}$$

Table 6.5 shows the annuity factors a for different interest periods n and interest rates ir. The annuity factor is sometimes simply called *annuity* and the interest rate is also known as the *discount rate*.

With the annuity factor a and the annual generated amount of energy E_a the levelled cost for one unit of energy c_E is easy to calculate:

$$c_E = \frac{B}{E_a} = \frac{c_0 \cdot a}{E_a} \tag{6.15}$$

The level of the interest rate depends on the risk of the investment. *Risks associated with renewable energy systems can include the overestimation of the available renewable energy resource such as solar irradiation or wind energy at one location, unforeseen technical troubles or changing legal conditions. Since these risks are usually higher than those of a typical bank account, the discount rates are also higher. However, it is nearly impossible to make reliable calculations for the level of the discount rate. In the end, the market with all its moods and the subjective feelings of investors define the discount rate.

The calculations above can be extended by applying price increases on operating and maintenance costs or fuel costs for conventional energy systems. However, statements on price increases over long investment periods are very difficult to make. Unforeseen events such as crises in oil-producing regions or shortages of fossil energy resources can change future fuel or raw material prices significantly. Therefore, statements on future fuel prices or operating and maintenance costs are not given here; instead the final sections give a critical overview of conventional interest calculations.

Costs for a photovoltaic system with return on capital

For the example of the photovoltaic system (A_0 = €6500, A_{10} = €1500, *ir* = 0.06, *q* = 1.06, *n* = 25, E_a = 800 kWh$_{el}$) of the section above, the calculations now become:

Discounted payments: c_0 = €6500 + €1500 · 1.06^{-10} = €7338

Annuity factor: $a = (1.06 - 1)/(1 - 1.06^{-25})$ = 0.0782

Required repayments p.a.: *B* = €7338 · 0.0782 = €574

Electricity generation costs: c_E = €574 / 800 kWh$_{el}$ = €0.72/kWh$_{el}$

When generating 2000 kWh of electricity per year in regions with very high annual solar irradiation, the electricity generation cost decreases to €0.29/kWh$_{el}$. However, the electricity generation cost including a return on capital is much higher than the cost without return on capital, as the above example illustrates (€0.72/kWh$_{el}$ compared with €0.40/kWh$_{el}$).

Costs for a wind power plant with return on capital

Adapting the example of the 1500 kW wind power plant (A_0 = €1,800,000; A_i = €50,000; *q* = 1.08; *n* = 20; E_a = 3.5 · 10^6 kWh$_{el}$) to an interest rate of *ir* = 8 per cent yields:

c_0 = €1,800,000 + €50,000 · 9.82 = €2,291,000

a = 0.1019

c_E = €2,291,000 · 0.1019 / 3.5·10^6 kWh$_{el}$ = €0.067/kWh$_{el}$

A high number of privately financed wind power projects have been built in the past few years in Germany. Many projects have been realized with 30 per cent equity capital. The remainder of the investment has come from bank loans with relatively low interest rates in the range of 5 per cent. Project risks such as incorrect yield calculations or changes in the wind resources are borne by the equity investor. Therefore, higher interest rates are assumed here.

Costs for a solar thermal system with return on capital

For the cost calculations of solar thermal systems, the numeric values of the calculations without return on capital are again used. With an interest rate of 6 per cent and an operating period of 15 or 20 years the annuity factor becomes 0.1030 or 0.0872, respectively. Table 6.6 shows the heat generation costs for different investment costs, operating periods and annual substituted amounts of energy. Most solar heat production costs can only compete with electrical water heating systems. To compete with gas or oil heating systems they usually need public subsidies at present. In regions with high solar irradiations and low labour and investment costs, solar energy systems can also compete with fossil-fired heating systems.

Table 6.6 *Levelled Heat Costs in €/kWh$_{therm}$ for Solar Thermal Systems for Domestic Water Heating with an Interest Rate of 6 per cent and with Annual Operating Costs of €40*

Substituted amount of energy	Annual 2300 kWh$_{therm}$		Annual 1150 kWh$_{therm}$	
Lifetime	20 years	15 years	20 years	15 years
Investment cost				
€5000	0.21	0.24	0.41	0.48
€3500	0.15	0.17	0.30	0.35
€2500	0.11	0.13	0.22	0.26
€1800	0.09	0.10	0.17	0.20
€1000	0.06	0.06	0.11	0.12

Future development of costs for renewable energy systems

Costs for renewable energy systems will decrease further as they have done in the past. Increased production volume, more automation in production and the use of ever more sophisticated technologies will reduce the costs significantly. Production volumes of many renewable energy technologies are still relatively low and many involve multiple production steps, requiring expensive labour.

An important parameter for future cost estimations is the so-called *progress ratio PR*. This parameter expresses the rate at which costs decline each time the cumulative capacity implementation doubles. For instance, a PR of 90 per cent corresponds to a learning rate of 10 per cent, i.e. there is a 10 per cent cost reduction for each doubling of the cumulative capacity.

The PR of *wind power plants* for the 1980s and 1990s was between 0.8 and 0.96 depending on the region (Harmsen and van Sambeek, 2003; IEA, 2000). The global installed capacity has doubled approximately every 2.5 years over the past decade. With an average PR of 0.92 the cost reduction is nearly 30 per cent per decade. Further cost reductions can be expected by increasing system sizes. However, in some regions, such as locations in Germany, wind turbines have been already installed at most of the best sites. Using further sites with lower wind speeds can offset part of the cost reduction of increased wind turbine production.

Photovoltaic systems have also achieved noteworthy cost reductions over the past few decades. Figure 6.3 shows the reduction in photovoltaic module prices for three countries. The global progress ratio for photovoltaic systems has been around 0.8 over the past few decades. The installed global photovoltaic capacity has been doubling approximately every three years. The resulting cost reduction is about 50 per cent per decade. Further significant cost reductions are possible due to new solar cell materials and fast-increasing production volumes. Most other renewables also show progress ratios of between 0.8 and 0.9.

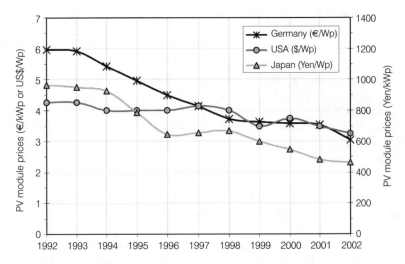

Source: IEA-PVPS, 2003

Figure 6.3 *Photovoltaic Module Prices in Germany, Japan and the USA*

This chapter will not give a prediction for future cost developments, because so many unknown parameters will influence them. However, if future progress ratios are in the same range as the progress ratios of the past few decades, renewable energy systems will surely become competitive with all conventional energy types within the 21st century. Besides cost reduction of renewable energy systems, the increase in fuel costs for conventional systems due to limited conventional energy resources will force this development in the long term.

However, exactly when renewable energy systems will dominate the global energy sector depends mainly on the question of when our society will give these energy resources a high priority to counteract the negative consequences of using conventional energy sources.

Costs of conventional energy systems

Conventional systems are often compared to renewable energy systems with respect to economics. However, external costs are often excluded from this comparison. Therefore, the following sections will discuss this point in more detail. Without considering external costs, the *levelled electricity generation costs* of big power stations are in the range of €0.03/kWh$_{el}$ to €0.07/kWh$_{el}$ for coal-fired power plants and between €0.03/kWh$_{el}$ and €0.04/kWh$_{el}$ for natural gas-fired combined gas and steam turbine power plants.

Private gas or oil heating systems for space heating usually have heat generation costs of between €0.05 and €0.06/kWh$_{therm}$. The heat generation costs for domestic water heating are normally a little higher because the conventional heating systems operate at part-load with a lower efficiency

Table 6.7 *Average Energy Prices in Germany for 2001*

Energy carrier and consumer groups	€/MWh	Energy carrier and consumer groups	€/MWh
Imported hard coal (free frontier)	6.5	Natural gas (public utility)	17.3
German hard coal ex colliery[c]	18.3	Natural gas (industry)	21.2
Crude oil (free frontier)	17.3	Electricity (industry)[a]	63.9
Domestic fuel oil (inland, industry)	28.9	Electricity (private household)[b]	143.0

Note: a without electricity tax and VAT; b all duties included; c costs for 1998
Source: Quaschning, 2003

during summer, when they only heat up tap water. The heat generation costs of electric water heating systems are above the electricity price that is about €0.14/kWh$_{el}$ for end users.

Not all electricity or heat generation costs of systems using renewable energy sources can be compared with conventional systems. Systems for solar domestic water heating usually need a fossil backup, in other words, the investment costs for the conventional system must be paid anyway. This means that the solar energy system can save only costs for the fuel of the conventional system and in some cases reduce maintenance costs and increase lifetime of the conventional system. Table 6.7 shows fuel costs in Germany; however, these costs will vary from country to country.

The electricity and heat generation costs given assume stable costs for fossil fuels; however, that will certainly not be the case for the long operating periods in the power plant sector. Figure 6.4 shows the changes in crude oil prices over the past three decades. If they are adjusted for inflation and exchange rate, the maximum of the past 30 years is 5.5 times higher than the minimum.

Prices for conventional energy carriers will increase in the long term due to the limited availability of fossil energy resources. High oil prices as already seen in the 1980s could occur again within the near future. Then, today's relatively expensive renewable energy systems will suddenly be more competitive.

EXTERNAL COSTS

Electricity and heat generation costs of conventional energy systems such as fossil or nuclear power plants are also calculated with the formulae described above. Besides the investment costs for the power plant, fuel costs and operation and maintenance costs, there are further external costs that may be only partly covered or not covered at all by the plant operators.

Besides government subsidies and government-financed research, development and disposal, external costs mainly comprise costs for damage to the environment and public health caused by the power plant. The electricity or heat price usually does not contain these external costs. This distorts the

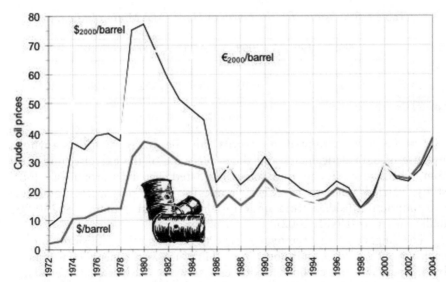

Figure 6.4 *Crude Oil Prices Given in Actual Prices and Adjusted for Inflation and Exchange Rate (Shown in $, $$_{2000}$ and €$_{2000}$, 1 barrel = 158.7579 litre)*

competition between renewable and conventional energy systems, because external costs of renewable energy systems are usually much lower. However, the estimation of all external costs is very difficult and often highly controversial, because many assumptions have to be made.

Subsidies in the energy market

There is much confusion about what is meant by an energy subsidy. The narrowest and most often used definition is a direct cash payment by a government to an energy producer or consumer. But this is just one possibility for stimulating production or use of a particular fuel or form of energy. The International Energy Agency (IEA) defines energy subsidies as any government action which concerns primarily the energy sector and which lowers the cost of energy production, lowers the price paid by energy consumers or raises the price received by energy producers. In practice, all energy subsidies are justified by one or more of the following reasons (UNEP, 2001):

- to protect employment in a particular indigenous industry or sector against international competition or to promote job creation
- to stimulate regional or rural economic development
- to reduce dependence on imports for energy-security reasons
- to lower the effective cost of and/or provide access to modern energy services for specific social groups or rural communities as a means of welfare support
- to protect the environment.

Table 6.8 *Subsidies for the German Hard Coal Mining Industry*

Year	1996	1997	1998	1999	2000	2001	2002
Subsidy (million €)	5059	4637	4510	4308	3972	3696	3049

Source: Quaschning, 2003

Few studies have attempted to quantify subsidies for the world as a whole, because of data deficiencies and the sheer scale of the exercise, and most comprehensive studies are now somewhat dated. The most prominent global study, carried out by the World Bank in 1992, put world fossil-fuel consumption subsidies from under-pricing alone at around US$230,000 million per year (UNEP, 2002).

The impact of the removal of energy consumption subsidies on the environment would be enormous. It would reduce world energy consumption by 3.5 per cent and carbon dioxide emissions by 4.6 per cent. Very large subsidies exist in Russia, China and India. In Iran the average rate of subsidy of the market price is about 80 per cent.

Although the highest subsidies exist in developing countries, fossil-fuel subsidies in highly industrialized countries such as Germany are also noticeable. Table 6.8 shows the subsidies for the German hard coal mining industries. Up to the year 2010, the total cumulated subsidies in this sector will be more than €80,000 million. This amount would have been sufficient to install nearly 70,000 MW of wind power, which could generate about one-third of the German electricity demand.

Over recent years the subsidies for renewable energy technologies have been increased. However, they are still far behind the enormous subsidies given to the fossil and nuclear energy sector. The high subsidies for conventional types of energy seriously interfere with climatic protection and the market for the introduction of renewables.

Expenditure on research and development

In many industrialized countries the majority of the expenditure on energy research and development (R&D) has been spent on nuclear power during the past few decades. Even today, the highest budgets are allocated for nuclear

Table 6.9 *Expenditure of the German Government on Energy Research and Development in Millions of Euros*

Period	1956–1988	1989–2002
Coal and other fossil energy sources	1907	340
Nuclear power	18,855	5384
Renewables and conservation	1577	1897

Source: Quaschning, 2003

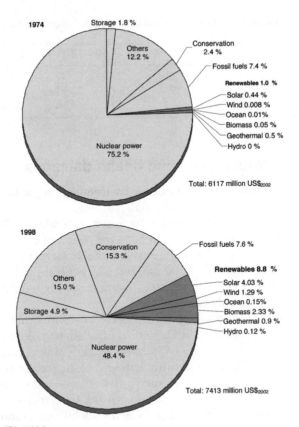

Source: data from IEA, 2003

Figure 6.5 *IEA Total Reported Government Energy Technology R&D Budgets for 1974 and 1998*

power, both nuclear fusion and fission. Table 6.9 shows the R&D expenditure of the German government since 1954 as an example.

The situation is similar in many other IEA countries. Figure 6.5 shows the IEA total reported government energy R&D budgets of the years 1974 and 1998. The majority of the R&D expenditure of the past few decades has also been spent on nuclear energy in these statistics. Even today, nearly half of the total R&D budget is allocated for nuclear energy. Before the mid-1970s there was almost no R&D budget for renewables. Since the oil crises and the Chernobyl nuclear accident, renewable energy budgets have increased; however, they were still rather low compared to the nuclear energy budgets at the end of the 1990s.

This unequal research policy causes a distortion of competition, mainly to the benefit of nuclear power. If the enormous R&D budgets for nuclear power had been spent on renewables, their costs would be much lower today and they could very probably compete in the global energy market without any other further subsidies.

Table 6.10 *Natural Disasters and Economic Losses*

Decade	1950–59	1960–69	1970–79	1980–89	1990–99
Number	20	27	47	63	89
Economic losses in billion US$$_{2001}$	42.2	75.7	136.1	211.3	652.3
Insured losses in billion US$$_{2001}$	---	7.2	12.4	26.4	123.2

Source: Munich Re, 2003

Costs for environmental and health damages

The estimation of costs of damage to the environment or public health is rather difficult and therefore very controversial.

The consequential damages of a maximum credible accident in a central European nuclear power plant would be in the region of €5000 billion. This is much more than a national economy can handle. In Germany the legal liability for a nuclear reactor is only €500 million. The general public would have to carry the remaining costs. Unlimited insurance cover would make nuclear power plants unprofitable in most cases.

Nuclear power stations can also cause costs in other countries. For instance, the reactor accident at Chernobyl affected many European countries, notably, through radioactively contaminated agricultural products. Yet it is not only reactor accidents that cause external costs; the normal operation of nuclear power plants also causes such costs. The consequential damages of uranium mining are enormous: damages arising from former uranium mines in Eastern Germany are estimated to exceed €6.5 billion. Uranium processing, enrichment, operation of power plants, transportation and final storage also cause external costs that are not currently paid by power plant operators.

The use of fossil energy resources also causes high indirect costs that are not covered by the energy price. *Damage to buildings*, *destruction of forests* or *damage to health* due to air pollution are only some obvious examples. Material damages alone in Germany due to air pollution caused by the use of fossil energy resources are estimated at more than €2 billion per year. Costs of damage to health such as respiratory tract diseases, allergies or cancers caused by fossil energies are difficult to estimate. They may also be in the range of several billion Euros per year.

The cost of *damage caused by global warming* cannot be estimated today. International insurance companies claim that there was a large increase in the occurrence of natural disasters over recent decades (see Table 6.10). Whether this increase was caused by the anthropogenic greenhouse effect cannot be proven today with absolute certainty. However, these numbers indicate what immense costs can arise from the greenhouse effect.

Other external costs

When using nuclear power there are also high costs for radioactive waste disposal. Besides a high amount of capital for exploitation and installation of possible final disposal sites there are other costs. Protests against transportation of highly radioactive waste in Germany in 1997 caused the most extended police operation of the past 50 years. The cost of the 30,000 policeman needed to protect the transport was about €50 million.

Further external costs are difficult to estimate. These are costs for administrative acts, licensing procedures and installation of infrastructure for power plants. The public radioactivity measurement system network and the public disaster protection make up other external costs.

Internalization of external costs

The considerations above show clearly that the external costs of renewable energy systems are often much lower than those of fossil or nuclear power plants. To alleviate these costs, the conventional energy sources could be made to pay compensatory levies, which could be used to repair the damage and to convert the energy supply to renewable energy sources with lower external costs.

The quantification of external costs and compensatory levies is not easy. On one hand, much environmental damage is not clearly assigned to a single polluter and many consequences are not yet known today. Many studies on this subject have been published, some of them with very contradictory results.

Hohmeyer carried out one of the first extended examinations (Hohmeyer and Ottinger, 1991). Besides costs for quantifiable damages to the environment he considered costs for exploiting fossil and nuclear energy resources. Since fossil and nuclear energy resources will be consumed by a few generations of humanity, future generations will not be able to use them any more. Therefore, financial reserves must be created to compensate for the higher energy costs of the future. He also included costs for public goods, services, subsidies and R&D. He excluded psycho-social follow-up costs of diseases and deaths and indirect environmental effects, environmental follow-up costs of the nuclear fuel cycle, hidden subsidies and costs of the greenhouse effect.

He determined the total external costs of nuclear power as up to €0.36/kWh$_{el}$ (base year 1982). The external costs of fossil energies are a little lower. For the *current electricity supply* in Germany with a combination of fossil and nuclear power, he estimated average external costs of between

Table 6.11 *External Cost Figures for Electricity Production in the EU for Existing Technologies*

Energy source/ technology	External costs in €/kWh	Energy source/ technology	External costs in €/kWh
Coal	0.02–0.15	Biomass	0–0.03
Oil	0.03–0.11	Hydroelectricity	0.0003–0.01
Natural gas	0.01–0.04	Photovoltaics	0.006
Nuclear power	0.002–0.007	Wind	0–0.0025

Source: data from the European Commission, 2003

€0.026/kWh$_{el}$ and €0.133/kWh$_{el}$. The power plant operators should pay these costs when selling electricity to compensate for the external costs. However, this would almost double the current electricity price. The external costs of renewables are much lower. Neglecting external costs puts renewable energy systems at a significant disadvantage. Therefore, renewable energy resources are not used in a way that would be optimal for the long term national economy and society in general.

A recent study by the European Commission estimated the external costs for electricity generation in Europe (European Commission, 2003). The external costs for fossil power plants are in the same range as those of the Hohmeyer study. However, the external costs for nuclear power are estimated to be much lower by the EC due to very different assumptions about accident risks and exploitation of resources. This indicates not only the controversy that can arise when estimating external costs but also the much lower costs of renewables and the adverse competition conditions created if external costs are neglected.

CRITICAL VIEW OF ECONOMIC CALCULATIONS

Several methods exist for the estimation of costs. Usually, economic models are applied that include interest but no external costs. The previous sections have already described the limits of such classical economic calculations. Even if external costs are not considered to be important, there are some other aspects for a critical view on classical methods for economic calculations.

Infinite increase of capital

A *numerical example* of Goetzberger is discussed here (Goetzberger, 1994). In this example, one Eurocent is invested at an interest rate of 4 per cent at the birth of Christ in the year 1 AD. The compound interest formula estimates what this investment would be worth in the year 2000. With

$$c_{2000} = €0.01 \cdot (1+0.04)^{2000} = €1.166 \cdot 10^{32}$$

the result would be an incredible €1.166·10^{32}. With a gold price of €10,500/kg this amount corresponds to 1.1·10^{28} kg of gold. The density of gold is 19.29 kg/dm^3 so that the gold would have a volume of 5.8·10^{14} km^3. For comparison, the volume of the earth is only 1.1·10^{12} km^3. In other words, the volume of gold would be 500 times greater than the volume of the earth.

This example shows clearly that the compound interest formula does not work for long periods of time. A *continuous growth of capital* is not possible over long periods of time because the compound interest formula *becomes infinite*. Nobody on Earth would be able to carry the interest burdens. Hence, all classical economical calculations that use the compound interest formula do not work over long periods of time. If these economical models are to be used, the question of over what period do they produce practical results must first be answered. However, it is difficult to define the longest period of time that is applicable for compound interest calculations. For instance, at an interest rate of 8 per cent, the invested capital increases tenfold within 30 years. In 100 years the invested capital has increased 2200 times. These periods are already relevant for investments in the energy sector.

With increasing periods of time and increasing interest rates, the probability of losing part of the capital, or even the whole capital, increases significantly. In the past, wars, currency reforms and environmental catastrophes have caused the total loss of available capital. In the long term, these events are actually necessary to keep our economic system running, otherwise the compound interest would bring infinite consumption in which case, the interest rate would become zero. Nowadays many people have difficulty in accepting that large-scale capital losses could also happen to them since the past 50 years have been relatively stable. In the first half of the 20th century, however, events such as the two world wars and the Great Depression destroyed huge amounts of capital. In the near future, natural disasters will be more and more responsible for capital losses (see section on costs for environmental and health damages, p251).

Sometimes, an investment even causes the events that destroy the investment. An example is a war loan in the case where the war is lost. At the end, multiples of the invested capital are lost. The same could happen in the energy sector. An investment in a nuclear power plant can not only cause the loss of the invested capital but also the loss of much greater assets if a maximum credible accident occurs. Investments in fossil energies support the greenhouse effect; the consequences of the greenhouse effect such as huge storms or floods will also cause high losses of capital. However, these events will not usually destroy the fossil energy installations that have been financed with the investment.

Increasingly, arguments are put forward that high capital growth should be sacrificed for the protection of existing capital. Investments in technologies that avoid the negative effects of other technologies can help in securing capital. Investment in renewable energy systems should also be seen from this point of view.

Responsibility of capital

It is nearly impossible for an investor to realize all the possible consequences of an investment. However, the higher the rates of return the higher the risks; this is a well-known rule. The risk for investments in the energy sector does not only include the invested capital but also much greater assets as well. Certainly, not all investments that receive a return on capital will eventually cause losses in capital, but all investments do have responsibilities, especially in the energy sector. Article 14 of the constitution of Germany expresses this appropriately:

> 'Property imposes duties. Its use should also serve the public weal.' (Tschentscher, 2003)

The question whether investments in renewable energies should have to yield a return needs to be asked. When buying a luxury car, few people think about profitability. Indeed, there are other economic options available to transport people from one site to another. However, tens of thousands or even hundreds of thousands of Euros can be spent on a luxury car without the investor asking for a rate of return. Yet if the same amount of money were to be invested in renewable energy systems, the investor would demand a return on capital. Arguments in favour of investing in a car include improved comfort and a higher quality of life; however, are these arguments not also valid for renewable energy systems? The use of energy resources that have a lower negative impact on nature and public health surely increases the quality of life of many people. However, the people who benefit from these investments are not necessarily the same as those who make the investment.

Many individuals have already installed solar energy systems without focusing on getting a good return on their investment. Some companies have even installed unprofitable systems, citing an improvement in corporate image as the reason for the investment. A gain in prestige is a concept that is difficult to express in economic terms. Therefore, it is to be hoped that architectonically presentable solar energy systems will in the future achieve the same sentimental value as a luxury car or a fur coat does today.

If all the points discussed above are considered, renewable energy sources should achieve a significantly higher share of the energy supply market very soon to minimize the risks of global warming or nuclear accidents. Only then will it be possible to keep our planet in a condition that can support a life worth living for future generations.

Chapter 7

Simulations and the CD-ROM of the Book

INTRODUCTION TO COMPUTER SIMULATIONS

The importance of computer programs for simulating and analysing renewable energy systems has increased during recent years. These programs are often used for predicting the energy gain, defining and optimizing systems or conducting economical calculations. Simulation software is also of importance for developing new system concepts. Analysing and simulating systems before building and commissioning can avoid problems, which could occur when deploying a prototype. Hence, simulation can save time and cost at the development stage as well as acting as quality control in the operational phase.

The number and quality of computer programs available have increased during recent years, alongside the increasing importance of renewable energy systems. Besides established software packages that have been in use for many years, there are always interesting new developments. Software in the marketplace covers almost all renewable applications possible today. This chapter aims to give access to available software; it is not a manual on how to implement the scientific descriptions given in the remainder of this book. The aim is to advise on suitable software packages, which can be found on the included complementary CD-ROM, either as full versions or as demonstration versions. A description of the software is not given in this book, as this would go beyond the scope of this chapter. However, detailed descriptions and manuals for programs are included on the CD-ROM.

The diversity of available professional programs does not supersede the theory of the renewable energy systems described in this book. Simulation results are only as good as the algorithms used. Only the understanding of the algorithms used allows a judgement to be made of the quality of any program and its results. However, the algorithms underlying the simulations are not always described for simulation programs. In most cases, the user is expected to trust the results blindly. The only possibility for verifying the results is a knowledge of basic calculation methods, which are described in the relevant chapters. Simulation results should always be validated when using computer software. A good feeling for the plausibility of the results is crucial. Although it is not possible to check all calculations in detail, some key results such as annual energy yield or specific costs should be compared with similar numbers

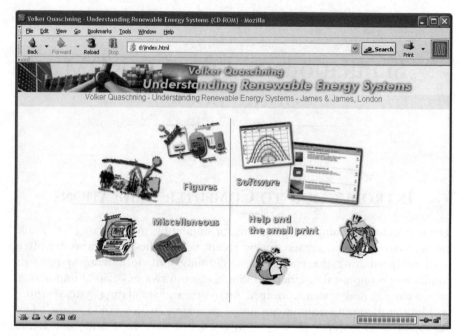

Figure 7.1 *Start Screen of the CD-ROM of the Book (Presentation with Mozilla Browser)*

from systems in operation. Sometimes, performing the simulation with two different computer programs is useful. If both results are of the same order of magnitude, the only mistakes possible are a result of badly chosen input parameters or unverifiable assumptions.

THE CD-ROM OF THE BOOK

Getting started and overview

The user-friendly navigation through the contents of the CD-ROM is achieved in HTML and an autostart option. Common browsers such as Mozilla, Netscape or the Microsoft Internet Explorer are suitable for the correct presentation. If the start screen does not appear automatically after inserting the CD-ROM into the CD drive, the file *index.html* in the main directory of the CD should be started with the browser. This initiates the start screen that is shown in Figure 7.1. The contents are split into *Figures, Software, Miscellaneous* and *Help*. Before using the CD, reading the *Help* section is recommended.

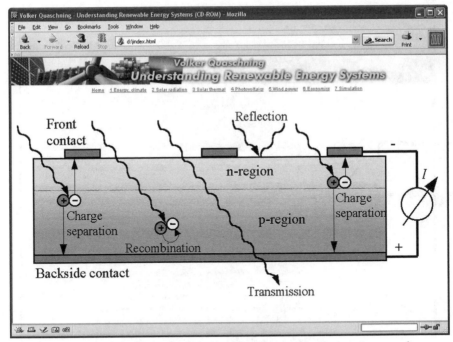

Figure 7.2 *All Figures are Included and Can be Chosen Separately*

Figures

The *Figures* section contains all the figures of the book in an electronic format, either GIF or JPG. Any figure can be selected via a user-friendly interface. The browser displays the figure as shown in Figure 7.2 for a figure from the chapter on photovoltaics. When using the figures, copyright law must be considered. The section entitled *Help and the small print*, which can be started from the main screen, gives further information.

Software

The *software* section can be accessed directly from the start screen. This section contains an extensive collection of computer programs dealing with renewable energy sources. The programs are subdivided into the fields of solar radiation, low-temperature solar thermal systems, high-temperature solar thermal systems, photovoltaics, wind power, education and economics. An alphabetical list, which is shown in Figure 7.3, gives an overview of all the programs on the CD-ROM. After selecting the software required, a short description appears with application area, hardware requirements and a link to the author or retailer.

All programs can be started or installed directly from the short program description. A detailed description or manual in the PDF format is added for some programs. These can be also started from the program description page. Some browsers do not allow the starting of programs directly. In this case, the program must be started using Windows Explorer. Therefore, the path required and the file to be started is also given in the program description.

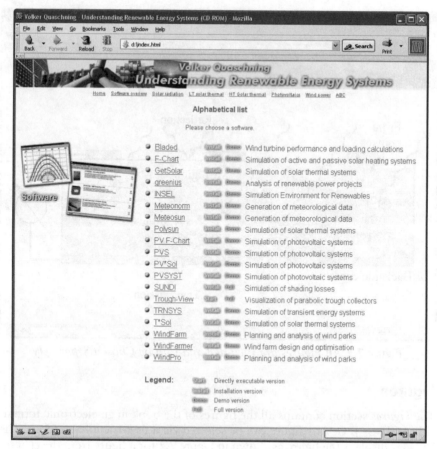

Figure 7.3 *Alphabetical Overview of all Software Programs on the* CD-ROM

Miscellaneous

The *Miscellaneous* section contains further text files in PDF format and program source codes. In addition to the details included in Chapter 2 of this book, some algorithms for the calculation of the position of the sun are added as source code in the computer language C or Delphi. This source code can be displayed with the browser.

Appendix

NATURAL CONSTANTS

Electron rest mass	m_e	$= 9.1093897 \cdot 10^{-31}$ kg
Proton rest mass	m_p	$= 1.6726231 \cdot 10^{-27}$ kg
Atomic mass unit	u	$= 1.660565 \cdot 10^{-27}$ kg
Elementary electron charge	e	$= 1.60217733 \cdot 10^{-19}$ A s
Permeability of free space	μ_0	$= 4p \cdot 10^{-7}$ V s /(A m)
Permittivity	ε_0	$= 8.85418781762 \cdot 10^{-12}$ A s/(V m)
Boltzmann constant	k	$= 1.380658 \cdot 10^{-23}$ J/K
Stefan–Boltzmann constant	σ	$= 5.67051 \cdot 10^{-8}$ W/(m^2 K^4)
Speed of light in a vacuum	c	$= 2.99792458 \cdot 10^8$ m/s
Planck's constant	h	$= 6.6260755 \cdot 10^{-34}$ J s
Solar constant	E_0	$= 1367 \pm 2$ W/m^2
Surface temperature of the sun	T_{Sun}	$= 5777$ K
Specific emission of the sun	$M_{e,S}$	$= 63.11$ MW/m^2

UNITS SOMETIMES USED IN THE US AND UK

Length	1 ft (foot) = 12 in (inch) = 1/3 yd (yard) = 0.3048 m
Length	1 mile = 1760 yd = 1609.344 m
Mass	1 lb (pound) = 16 oz (ounce) = 0.45359237 kg
Volume	1 gal (US gallon) = 231 in^3 = 3.7854345 litre
Volume (crude oil)	1 ptr.bbl (petroleum barrel) = 42 ptr.gal (petroleum gallon) = 158.7579 litre
Temperature	ϑ_F (in °F, degrees Fahrenheit) = 9/5 · ϑ (in °C, degrees Celsius) + 32
Power	1 h.p. (horse-power) = 745.7 W
Pressure	1 inHg (inch column of mercury) = 25.4 mmHg = 3386.3788 Pa
Energy	1 BTU (British thermal unit) = 1.05505585262 kJ = 0.000293071 kWh
Irradiation	1 ly (langley) = 1 cal/cm^2 = 4.1868·104 J/m^2 = 0.01163 kWh/m^2

References

CHAPTER 1

Blasing, T. J.; Jones, S. (2003) Current Greenhouse Gas Concentration. Carbon Dioxide Information Analysis Center, available at *http://cdiac.esd.ornl.gov*

BGR, Federal Institute for Geosciences and Natural Resources (2002) *Reserves, Resources and Availability of Energy Resources 2002*. Berlin, German Federal Ministry of Economics and Labour

British Petroleum (2003) *BP Statistical Review of World Energy 2003*. London, BP

DOE, US Department of Energy (2003) *International Energy Annual*. Washington DC, DOE

Ender, C. (2004) Wind Energy Use in Germany 2004. *DEWI Magazin*, no 24, pp6–18

Enquete-Kommission (1995) 'Schutz der Erdatmosphäre' des 12. Deutschen Bundestages: *Mehr Zukunft für die Erde*. Bonn, Economica Verlag

European Solar Thermal Industry Federation, ESTIF (2003) *Sun in Action II – A Solar Thermal Strategy for Europe*. Brussels, ESTIF

International Energy Agency, IEA (2003a) *Key World Energy Statistics*. Paris, IEA

International Energy Agency, IEA (2003b) *Renewable Information 2003*. Paris, IEA

International Energy Agency – PV Power Systems Programme, IEA-PVPS (2003) *Trends in Photovoltaic Applications between 1992 and 2003*. Report IEA-PVPS T1-12.

Intergovernmental Panel on Climate Change, IPCC (2000) *IPCC Special Report Emissions Scenarios*. Nairobi, IPCC

Intergovernmental Panel on Climate Change, IPCC (2001) *Third Assessment Report of Working Group I, Summary for Policy Makers*. Shanghai, IPCC

Itaipu Binacional (2003) *Itaipu Binational Technical Data*, available at *http://www.itaipu.gov.br*

Quaschning, V. (2001) Hydrogen to Meet the Storage Demand for a Future Climate-Compatible Electricity Supply in Germany. *Proceedings of Hypothesis IV Symposium*. Stralsund, 9–14 September 2001, pp113–117

United Nation Framework Convention on Climate Change, UNFCCC (1998) *Methodological Issues while Processing Second National Communications: Greenhouse Gas Inventories*. Buenos Aires, FCCC/SBSTA

United Nation Framework Convention on Climate Change, UNFCCC (2002) *Report on National Greenhouse Gas Inventory Data from Annex I Parties for 1990 to 2000*. New Delhi, FCCC/SB/2002/INF.2

Voss, K.; Goetzberger, A.; Bopp, G.; Häberle, A.; Heinzel, A.; Lehmberg, H. (1995) The Self-Sufficient Solar House Freiburg – Results of Three Years of Operation. *Proceedings of ISES Solar World Congress*, Harare, 11–15 September

V. Weizsäcker, E. U.; Lovins, A. B.; Lovins, L. H. (1998) *Factor Four: Doubling Wealth – Halving Resource Use: A Report to the Club of Rome*. London, Kogan Page

CHAPTER 2

CEN, European Committee for Standardization (1999) *EN ISO 9488: Solar Energy Vocabulary*. Brussels, CEN

Deutsches Institut für Normung e.V., DIN (1982) *DIN 5031, Optical Radiation Physics and Illuminating Engineering*. Berlin, Beuth Press

Deutsches Institut für Normung e.V., DIN (1985) *DIN 5034 Part 2, Daylight in Indoor Rooms.* Berlin, Beuth Press

Dietze, G. (1957) *Einführung in die Optik der Atmosphäre.* Leipzig, Akademische Verlagsgesellschaft Geest & Portig K.G.

International Solar Energy Society, ISES (2001) Units and Symbols in Solar Energy. Solar Energy Journal vol 71, ppIII–V

International Organization for Standardization, ISO (1993) *ISO Standards Handbook Quantities and Units.* Geneva, ISO

Kambezidis, H. D.; Papanikolaou, N. S. (1990) Solar Position and Atmospheric Refraction. *Solar Energy* vol 44, pp143–144

Klucher, T. M. (1979) Evaluation of Models to Predict Insolation on Tilted Surfaces. *Solar Energy* vol 23, pp111–114

NASA (2003) Surface Meteorology and Solar Energy, Release 4, available at *http://eosweb.larc.nasa.gov*

National Renewable Energy Laboratory, NREL (2000) Solpos 2.0 Documentation, available at *http://rredc.nrel.gov/solar/codesandalgorithms/solpos*

Palz, W.; Greif, J. (1996) *European Solar Radiation Atlas.* Berlin, Springer

Perez, R.; Stewart, R. (1986) Solar Irradiance Conversion Models. *Solar Cells* vol 18, pp213–222

Perez, R.; Seals, R.; Ineichen, P.; Stewart, R.; Menicucci, D. (1987) A New Simplified Version of the Perez Diffuse Irradiance Model for Tilted Surfaces. *Solar Energy* vol 39, pp221–231

Perez, R.; Ineichen, P.; Seals, R.; Michalsky, J.; Stewart, R. (1990) Modeling Daylight Availability and Irradiance Components from Direct and Global Irradiance. *Solar Energy* vol 44, pp271–289

Quaschning, V.; Hanitsch, R. (1998) Irradiance Calculations on Shaded Surfaces. *Solar Energy* vol 62, pp369–375

Quaschning, V.; Ortmanns, W. (2003) Specific Cost Development of Photovoltaic and Concentrated Solar Thermal Systems Depending on the Global Irradiation. *Proceedings of ISES Solar World Congress 2003,* 4–19 June 2003, Gothenburg

Reindl, D. T.; Beckman, W. A.; Duffie, J. A. (1989) Diffuse Fraction Correlations. *Proceedings of ISES Solar World Conference 1989,* 4–8 September, Kobe, pp2082–2086

Sattler, M. A.; Sharples, S. (1987) Field Measurements of the Transmission of Solar Radiation through Trees. *Proceedings of ISES Solar World Conference 1987,* 13–18 September, 1987, Hamburg, pp3846–3850

Schulze, R. (1970) *Strahlenklima der Erde.* Darmstadt, Steinkoff

Technischer Überwachungsverein TÜV-Rheinland (1984) *Atlas über die Sonnenstrahlung in Europa.* Cologne, TÜV

Walraven, R. (1978) Calculating the Position of the Sun. *Solar Energy* vol 20, pp393–397

Wilkinson, B. J. (1981) An Improved FORTRAN Program for the Rapid Calculation of the Solar Position. *Solar Energy* vol 27, pp67–68

CHAPTER 3

Deutsches Institut für Normung e.V., DIN (1996) *DIN EN 1057, Copper and Copper Alloys – Seamless, Round Copper Tubes for Water and Gas in Sanitary and Heating Applications.* Berlin, Beuth Press

Hahne, E.; Kübler, R. (1994) Monitoring and Simulation of the Thermal Performance of Solar Heated Outdoor Swimming Pools. *Solar Energy* vol 53, pp9–19

Khartchenko, N. (1998) *Advanced Energy Systems.* New York, Taylor and Francis Group

Kleemann, M.; Meliß, M. (1993) *Regenerative Energiequellen.* Berlin, Springer
Lien, A. G.; Hestenes, A. G.; Aschehoug, O. (1997) The Use of Transparent Insulation in Low Energy Dwellings in Cold Climates. *Solar Energy* vol 59, pp27–35
Ladener, H. (1995) *Solaranlagen.* Staufen, Ökobuch Verlag
Manz, H.; Egolf, P. W.; Suiter, P.; Goetzberger, A. (1997) TIM-PCM External Wall System for Solar Space Heating and Daylighting. *Solar Energy* vol 61, pp369–379
Smith, C. C.; Löf, G.; Jones, R. (1994) Measurement and Analysis of Evaporation from an Inactive Outdoor Swimming Pool. *Solar Energy* vol 53, pp3–7
SPF Institut für Solartechnik (2002) *SPF Info CD 2002 Thermal Solar Energy.* Rapperswil, SPF
TiNOX GmbH (2004) TiNOX Absorbers. Available at *http://www.tinox.com*
Verein Deutscher Ingenieure, VDI (1982) VDI 2067 Blatt 4. *Economic Calculation of Heat-consumption: Installation of Warm Water Supplies.* Düsseldorf, VDI press
Wagner & Co. (Hrsg.) (1995) *So baue ich eine Solaranlage, Technik, Planung und Montage.* Cölbe, Wagner & Co. Solartechnik GmbH

CHAPTER 4

Goetzberger, A.; Voß, B.; Knobloch, J. (1998) *Crystalline Silicon Solar Cells.* Hoboken, John Wiley & Sons
Gretsch, R. (1978) *Ein Beitrag zur Gestaltung der elektrischen Anlage in Kraftfahrzeugen.* Nürnberg-Erlangen, Habilitationsschrift
Green, M. A. (1994) *Silicon Solar Cells.* Sydney, UNSW
Hasyim, E. S.; Wenham, S. R.; Green, M. A. (1986) Shadow Tolerance of Modules Incorporating Integral Bypass Diode Solar Cells. *Solar Cells* vol 19, pp109–122
Lasnier, F.; Ang, T. G. (1990) *Photovoltaic Engineering Handbook.* Bristol, Hilger
Lechner, M. D. (1992) *Physikalisch-chemische Daten.* Berlin, Springer
Luque, A.; Heqedus, S. (2003) *Handbook of Photovoltaic Science and Engineering.* Hoboken, John Wiley & Sons
Marti, A.; Luque, A. (2003) *Next Generation Photovoltaics.* Bristol, Institute of Physics
Michel, M. (1992) *Leistungselektronik.* Berlin, Springer
Photon International (2001) Market survey on grid inverters. Aachen, *Photon International*, pp52–60
University of Oldenburg, Department of Energy and Semiconductor Research (1994) *INSEL Manual (Interactive Simulation of Renewable Energy Supply Systems).* Oldenburg, University of Oldenburg
Quaschning, V.; Hanitsch, R. (1996) Numerical Simulation of Current-Voltage Characteristics of Photovoltaic Systems with Shaded Solar Cells. *Solar Energy* vol 56, pp513–520
Wagemann, H.-G.; Eschrich, H. (1994) *Grundlagen der photovoltaischen Energiewandlung.* Stuttgart, Teubner
Wolf, M.; Noel, G. T.; Stirn, R. J. (1977) Investigation of the Double Exponential in the Current-Voltage Characteristics of Silicon Solar Cells. *IEEE Transactions on Electron Devices* vol ED-24, No 4, pp419–428

CHAPTER 5

Betz, A. (1926) *Windenergie und ihre Ausnutzung durch Windmühlen.* Staufen, Ökobuch, Unveränderter Nachdruck aus dem Jahr
Carstens, J.; Stiebler, M. (1997) Active Filters for Power-smoothing and Compensation of Reactive Power and Harmonics. *Proceedings of European Wind Energy Conference 1997,* 6–9 October 1997, Dublin

Christoffer, J.; Ulbricht-Eissing, M. (1989) *Die bodennahen Windverhältnisse in der Bundesrepublik Deutschland*. Offenbach, Deutscher Wetterdienst, Ber.Nr.147

Enercon (1997) *Technical Documentation of the Wind Turbines E30, E40 and E66*. Aurich, Enercon GmbH

Fitzgerald, A. E.; Kingsley, C.; Umans, S. D. (2002) *Electric Machinery*. Columbus, McGraw-Hill

Gasch, R.; Twele, J. (Eds) (2002) *Wind Power Plants*. London, James & James

Hau, E. (2000) *Wind Turbines: Fundamentals, Technologies, Application and Economics*. Berlin, Springer

Hering, E.; Martin, R.; Stohrer, M. (1992) *Physik für Ingenieure*. Düsseldorf, VDI-Verlag

Hindmarsh, J. (1995) *Electrical Machines and Their Applications*. Oxford, Butterworth-Heinemann

Molly, J.-P. (1990) *Windenergie*. Karlsruhe, C.F. Müller

Risø National Laboratory, Wind Energy and Atmospheric Physics Dept. (1987) *WAsP - Wind Atlas Analysis and Application Program*. Riskilde, Risø National Laboratory

Tacke Windenergie GmbH (1997) *Technical Documentation of the Wind Turbine TW600*. Salzbergen, Tacke

Troen, I.; Petersen, E. L. (1989) *European Wind Atlas*. Roskilde, Risø National Laboratory - ISBN 87-550-1482-8

Vereinigung Deutscher Elektrizitätswerke, VDEW (1994) *Technische Richtlinie - Parallelbetrieb von Eigenerzeugungsanlagen mit dem Mittelspannungsnetz des Elektrizitätsversorgungsunternehmens (EVU)*. Frankfurt, VWEW Verlag

VDI (1993) *VDI Heat Atlas*. Berlin, Springer

Vestas (1997) *Technical Documentation of the Wind Turbines V42 and V44*. Husum, Vestas Germany GmbH

CHAPTER 6

Becker, G.; Kiefer, K. (2000) Kostenreduzierung bei der Montage von PV-Anlagen. Tagungsband *15. Symposium Photovoltaische Solarenergie*. Banz, OTTI, pp408–412

European Solar Thermal Industry Federation, ESTIF (2003) *Sun in Action II – A Solar Thermal Strategy for Europe*. Brussels, ESTIF

European Commission (2003) *External Costs – Research Results on Socio-environmental Damages Due to Electricity and Transport*. Brussels, EC

Federal Reserve Bank of Minneapolis: Consumer Price Index since 1800 (2003), available at *http://minneapolisfed.org/research/data/us/calc/hist1800.cfm*

Goetzberger, A. (1994) Wirtschaftlichkeit - Ein neuer Blick in Bezug auf Solaranlagen. *Sonnenenergie* vol 4, pp3–5

Harmsen, R.; van Sambeek, E. J. W. (2003) *Learning Curves*. Petten (Netherlands), ECN, Report Nr. ECN-C—03-074/H

Hohmeyer, O.; Ottinger, R. L. (1991) *External Environmental Costs of Electric Power*. Berlin, Springer

International Energy Agency, IEA (2000) *Experience Curves for Energy Technology Policy*. Paris, OECD/IEA

International Energy Agency (2003) IEA Energy Technology R&D Statistics, available at *http:// http://www.iea.org/stats/files/rd.htm*

International Energy Agency – PV Power Systems Programme IEA-PVPS (2003) *Trends in Photovoltaic Applications between 1992 and 2003*. Report IEA-PVPS T1-12:2003.

Munich Re (2003) *Topics – Annual Review: Natural Catastrophes 2002*. Munich, RE

Quaschning, V. (2003) *Regenerative Energiesysteme*. Munich, Hanser

Tschentscher, A. (2003) *The Basic Law – The Constitution of the Federal Republic of Germany*. Würzburg, Jurisprudentia

United Nations Environment Programme, UNEP (2001) *Energy Subsidy Reform and Sustainable Development: Challenges for Policymakers*. Synthesis Report, Paris, UNEP/IEA

United Nations Environment Programme, UNEP (2002) *Reforming Energy Subsidies*. Oxford, United Nations Publications

Wind-Kraft Journal (1993) Was kosten Anlagen 1993. *Wind-Kraft Journal* vol 1, p31

Wind-Kraft Journal (1999) Was kosten Anlagen 1999. *Wind-Kraft Journal* vol 1, pp12–14

Index

WIND POWER
Paul Gipe

'Paul Gipe's Wind Power is a must for everybody who's involved in the wind energy sector – or wants to be involved in the future. The reader will get a comprehensive overview of one of the most important energy technologies to save the world's climate: wind energy'
SVEN TESKE, Renewable Energy Director, Greenpeace

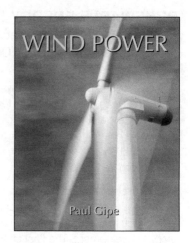

'This is THE definitive book on wind energy, beautifully, logically organized, with a great wealth of pictures, charts, graphs, formulas, cautionary tales and a lifetime of knowledge'
DOUG PRATT, Technical Editor, Real Goods

Wind Power is a completely revised and expanded edition of Paul Gipe's definitive 1993 book *Wind Power for Home and Business*. In addition to expanded sections on gauging wind resources and siting wind turbines, this edition includes new examples and case studies of successful wind systems, international sources for new and used equipment, and hundreds of colour photographs and illustrations.

Pb £35.00 1-902916-54-9
July 2004
Not available in the USA or Canada

SEND YOUR ORDERS TO:
James & James/Earthscan
FREEPOST NAT12094
8-12 Camden High Street
London
NW1 OYA
Tel: +44 (0)20 7387 8558
FOR MORE INFORMATION VISIT OUR WEBSITES:
www.earthscan.co.uk • www.jxj.com

WINNING THE OIL ENDGAME
Innovation for Profits, Jobs and Security
Amory B. Lovins, E. Kyle Datta, Odd-Even Bustness, Jonathan G. Koomey and Nathan J. Glasgow

'We can, as Amory Lovins and his colleagues show vividly, win the oil endgame... [A]n intriguing case that is important enough to merit careful attention by all of us, private citizens and business and political leaders alike'
GEORGE P. SHULTZ, Distinguished Fellow at the Hoover Institution, Stanford University, former Secretary of State, the Treasury, and Labor

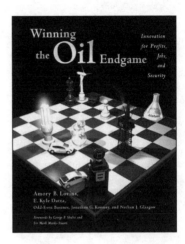

'*Winning the Oil Endgame* is a masterpiece in terms of stimulating ideas and discussion around how we move forward to realistically address our energy future. Rocky Mountain Institute has developed and analysed the data in a compelling way – worthy of much consideration'
BILL GLOVER, Director, Environmental Performance Strategy, Boeing Commercial Airplanes

ENOUGH ABOUT THE OIL PROBLEM. HERE'S THE SOLUTION.

Over a few decades, starting now, a vibrant US economy (then others) can completely phase out oil. This will save a net $70 billion a year, revitalize key industries and rural America, create a million jobs, and enhance security.

Here's the roadmap – independent, peer-reviewed, co-sponsored by the pentagon – for the transition beyond oil, let by business and profit.

Pb £49.50 1-84407-194-4
December 2004
Not available in the USA or Canada

SEND YOUR ORDERS TO:
James & James/Earthscan
FREEPOST NAT12094
8-12 Camden High Street
London
NW1 OYA
Tel: +44 (0)20 7387 8558
FOR MORE INFORMATION VISIT OUR WEBSITES:
www.earthscan.co.uk • www.jxj.com